Quantitative User Experience Research

Informing Product Decisions by Understanding Users at Scale

Chris Chapman
Kerry Rodden

Apress®

Quantitative User Experience Research: Informing Product Decisions by Understanding Users at Scale

Chris Chapman
Seattle, WA, USA

Kerry Rodden
San Francisco, CA, USA

ISBN-13 (pbk): 978-1-4842-9267-9
https://doi.org/10.1007/978-1-4842-9268-6

ISBN-13 (electronic): 978-1-4842-9268-6

Managing Director, Apress Media LLC: Welmoed Spahr
Acquisitions Editor: Jonathan Gennick
Development Editor: Laura Berendson
Editorial Assistant: Shaul Elson
Copy Editor: Bill McManus

Cover image by Ricardo Gomez Angel from Unsplash

Distributed to the book trade worldwide by Springer Science+Business Media LLC, 1 New York Plaza, Suite 4600, New York, NY 10004. Phone 1-800-SPRINGER, fax (201) 348-4505, e-mail orders-ny@springer-sbm.com, or visit www.springeronline.com. Apress Media, LLC is a California LLC and the sole member (owner) is Springer Science + Business Media Finance Inc (SSBM Finance Inc). SSBM Finance Inc is a **Delaware** corporation.

For information on translations, please e-mail booktranslations@springernature.com; for reprint, paperback, or audio rights, please e-mail bookpermissions@springernature.com.

Apress titles may be purchased in bulk for academic, corporate, or promotional use. eBook versions and licenses are also available for most titles. For more information, reference our Print and eBook Bulk Sales web page at http://www.apress.com/bulk-sales.

Any source code or other supplementary material referenced by the author in this book is available to readers on GitHub (https://github.com/Apress). For more detailed information, please visit http://www.apress.com/source-code.

Printed on acid-free paper

Table of Contents

About the Authors

Chris Chapman, PhD, is a Principal UX Researcher at Amazon Lab126, the founder and co-chair of the Quant UX Conference, and president of the Quantitative User Experience Association. Chris is the co-author of two popular Springer books on programming and analytics: *R for Marketing Research and Analytics* and *Python for Marketing Research and Analytics*. As a psychologist, Chris emphasizes the human focus of research and the need for integrated quantitative and qualitative understanding of users.

Kerry Rodden, PhD, is a Senior Principal Researcher at Code for America. Kerry founded the Quantitative UX Research role at Google in 2007 and managed the industry's first Quant UX research team. Kerry has originated popular tools and techniques, including the HEART metrics framework for user experience and the sequences sunburst visualization of user behavior. Kerry's background is in computer science and human-computer interaction, with a focus on the analysis and visualization of large-scale usage data, including A/B testing.

About the Technical Reviewers

James Alford, PhD, is a QUX professional with 15+ years of experience. He is the founder and principal investigator for Unabashed Research, LLC, a consulting company specializing in front-end product development research. From start to finish, he leads research projects designed to answer specific strategic business questions. He specializes in survey research that can be generalized to the larger market, and his work covers the full spectrum of product development, including early segmentation research, identifying groups of users with specific product needs, and conducting go-to-market research to price the perceived add value of new product features. Past clients include tech companies, online retailers, travel booking sites, job search sites, and more.

Katherine Joyce is a user experience professional, accessibility advocate, and design leader with over 9 years of experience having worked across a variety of sectors, including finance and government. She creates innovative, intuitive customer experiences and is an advocate of accessible design. As Head of Design at Capita Public Services, she is leading the UX vision, scaling multi-disciplinary design teams and supporting citizen-facing products driven by user needs. She is also a Design Mentor at DesignLab, where she provides professional global mentorship across a variety of courses combined with career development and portfolio advice for individuals who wish to pursue a career in product design. She also provides content, research, and design mentorship on ADPList, is a LUMA Institute certified Human-Centered Design Practitioner, and is Nielsen Norman Group Certified in the specialism UX Management.

Acknowledgments

This book has been in progress throughout the authors' entire careers—not as a defined project but in the course of learning, reflecting, and working with many others as the new discipline of quantitative user experience research has emerged. We have many people and institutions to thank, more than we are able to list here.

Special thanks go to colleagues who supported us in writing this book, especially Kitty Xu, co-founder of the Quant UX Conference. Kitty provided encouragement and feedback at every step, beginning with the initial concept, and read the entire manuscript multiple times.

We were also enormously fortunate to have had feedback from a superb group of researchers who were early readers: Tanesia Beverly, Regard Booy, Mario Callegaro, Maria Cipollone, Fei Gao, Joe Paxton, Mackenzie Sunday, Matt Stimpson, and Katie Wainwright.

At Apress, we thank our outstanding technical reviewers, Katherine Joyce and James Alford, and editors Shaul Elson, Jill Balzano, and Jonathan Gennick. Everyone at Apress improved this book while making the publication process smooth and productive.

Chris recognizes five leaders who introduced him to amazing UX research communities over the years: Paul Elrif, who hired him as a new-to-industry psychology postdoc to join Microsoft in 2000; Andy Cargile, who created a culture of growth and excellence at Microsoft Hardware; Lisa Kamm, who invited Chris to Google in New York, where he found lifelong friends; Dawn Shaikh, who led the Google Cloud UX team as it grew from a few UXers to a large, world-class organization; and Matt Roblee, whose leadership of the Amazon Lab126 industrial design research team is setting a new example of leading-edge quantitative and qualitative research.

Chris also gives thanks to a few favorite Pacific Northwest bookstores: Elliott Bay Book Company, Seattle; Powell's Books, Portland; University Book Store, Seattle; Griffin Bay Bookstore, Friday Harbor; Pariyatti, Onalaska; Holden Village Store, Holden; and Serendipity Books, Friday Harbor. Thank you to each one for your love of books and community.

ACKNOWLEDGMENTS

Kerry would like to thank the early Quant UXR team at Google for their willingness to take on and help define a brand new role: Aaron Sedley, Geoff Davis, Hilary Hutchinson, and Xin Fu. Leadership support from Maria Stone was essential to creating the role, and Jens Riegelsberger was a key collaborator in evolving and clarifying the hiring and performance review criteria.

Kerry is also grateful to others who contributed to the Quant UXR projects described in this book, or helped them to find a wider audience. In particular:

- The Google Ventures (GV) UX team's collective adoption and championing of the HEART framework (discussed in Chapter 7, "Metrics of User Experience") was essential to it becoming widely used outside of Google. Michael Leggett was the design lead on the Gmail labels project described in Chapter 7, and an early advocate for use of UX metrics by designers.

- Josh Sassoon suggested key design improvements to the sequences sunburst visualization in Chapter 9, "Log Sequence Visualization." Kent Russell's work on the sunburstR library made the visualization available to R users, and we are grateful to be able to use it in this book.

Finally, we thank our families, who were highly supportive and patient as we spent many weekends and evenings writing and editing over the past two years: Cristi and Maddie, and Liz and Theo.

Chris Chapman
Seattle, Washington
Friday Harbor, Washington

Kerry Rodden
San Francisco, California
March 2023

Introduction

Quantitative user experience research is a complex field and this book covers everything from the basic definition of *user experience* through several advanced examples of statistical analysis. To help orient you, here is a brief description of each chapter.

Part 1, "User Experience and Quant UX," sets the stage and orients you to our approach in this book and to user experience (UX) and quantitative UX roles.

- Chapter 1, "Getting Started," describes our intended audience and how we approach the role of quantitative UX in this book, where we try not to repeat other information sources but maximize unique information that will help you accelerate your understanding. This chapter provides pointers and help for later chapters that use data sets and R code.

- In Chapter 2, "User Experience and UX Research," we introduce the overall structure of user experience teams and their roles, including UX research.

- Chapter 3, "Quantitative UX Research: Overview," defines quantitative UX research and discusses how it differs from other UX research positions, data science, business analytics, and other roles.

Part 2, "Core Skills," describes the skills that are required for Quant UX research. UX researchers should be relatively expert in *human research* but they also need more modest skills in statistics and programming.

- Chapter 4, "UX Research," describes the unique approach to human-centered research that is a core of Quant UX.

- In Chapter 5, "Statistics," we examine the level of statistics knowledge that is required for Quant UX roles and projects.

- Chapter 6, "Programming," covers the degree of programming knowledge that we recommend and explains why basic programming is an important skill for Quant UX.

In Part 3, "Tools and Techniques," we review tools and techniques that are useful in Quant UX research. Our goal is not to describe everything that you might do, but rather to cover techniques and methods that are important and yet not covered well in other texts. The final three chapters in this section use data sets and R code to demonstrate how we analyze and interpret data in day-to-day practice.

- Chapter 7, "Metrics of User Experience," presents the HEART framework that helps teams to define user-centered metrics. It emphasizes multidimensional assessment using multiple data sources and perspectives on the user experience.

- Chapter 8, "Customer Satisfaction Surveys," gives guidance on writing surveys to assess user satisfaction and shares R code for basic analyses. We argue that satisfaction surveys should be as short as possible and describe several ways they commonly run into problems.

- In Chapter 9, "Log Sequence Visualization," we demonstrate how to analyze sequences of user behavior using sunburst charts. These are powerful exploratory tools to see how users behave over time in applications, on websites, or in similar sequential data.

- Chapter 10, "MaxDiff: Prioritizing Features and User Needs," reviews the MaxDiff survey method, which is useful to prioritize lists of features, needs, use cases, and product messages. This chapter also highlights the upper range of programming fluency that is helpful for Quant UX research.

Part 4, "Organizations and Careers," discusses how Quant UX researchers fit into product teams, and how they are hired, work with others, and build careers.

- Chapter 11, "UX Organizations," describes archetypal models for UX organizations along with their advantages and disadvantages.

- In Chapter 12, "Interviews and Job Postings," we review typical hiring processes for Quant UX research candidates and share recommendations on how to approach job interviews.

- Chapter 13, "Research Processes, Reporting, and Stakeholders," outlines how to work with key stakeholders in industry roles.

We emphasize the importance of focusing on product, design, and business *decisions* rather than on technical difficulty or accomplishment.

- Chapter 14, "Career Development for Quant UX Researchers," outlines career paths for Quant UX researchers. We highlight the importance of finding a fit between individual happiness and the demands of various roles and promotion levels. We propose three career models for very senior contributors.

- Finally, Chapter 15, "Future Directions for Quant UX," reflects on emerging trends and potential future directions for Quant UX.

The chapters are intended to stand alone and may be read in any order. They are connected with extensive cross-referencing of themes. In Chapter 1, we recommend several reading paths for different backgrounds and interests.

PART I

User Experience and Quant UX

Introduction to Part I

Part I begins with a discussion of this book's audience and approach in Chapter 1, "Getting Started." In Chapter 2, "User Experience and UX Research," we describe user experience (UX) roles, product development, research activities, and common questions.

If you are new to UX, we recommend to read all the chapters in Part I. For readers who already have UX experience, Chapter 3, "Quantitative UX Research: Overview," answers what is unique about Quant UX and compares it to other roles such as data science.

CHAPTER 1

Getting Started

Welcome! In this book, we present a complete introduction to quantitative user experience research, which we abbreviate as Quant UX research or Quant UXR. We intend this to be the single best book for you to learn about the practice of Quant UX research, the methods it uses, and whether a dedicated Quant UXR position might fit your interests.

Quant UXR is a complex term that is used to refer to two things: the *methods* of conducting quantitative UX research, and the *job role* of being a Quant UX researcher. In this book, we give attention to both. We emphasize the *role* in addition to the practice because Quant UX roles are a recent addition to the UX community, are growing rapidly, and are distinctive from other research positions. Despite that situation, there is no comparable text describing the work of Quant UXRs.

If you are not interested in Quant UXR roles—perhaps because you are a general UX researcher, data scientist, product manager, engineer, or designer—there are several chapters devoted to methods and frameworks that you could use. The chapters covering methods assume that you have technical background with statistics and programming, as described later in this chapter.

1.1 Who Are We? Why Should You Listen to Us?

We have been Quant UXRs for a combined total of more than 25 years, working at Amazon, Code for America, Google, and Microsoft.

Kerry has a PhD in computer science, specializing in human-computer interaction (HCI). While working at Google, Kerry became one of the first Quant UXRs, defined the formal role of Quant UXR, and was the first manager of *all* the Quant UXRs across Google. Some of Kerry's early Quant UX projects were reusable tools and techniques that were adopted by other organizations, and are represented in this book (Chapter 7, "Metrics of User Experience," and Chapter 9, "Log Sequence Visualization").

Chris has a PhD in psychology, and worked for 11 years as a general UX re searcher at Microsoft. He then joined Google, where he was a Quant UXR for more than 10 years and served on the company-wide hiring committee for UXRs. Currently Chris is a principal researcher at Amazon Lab126. He is the founding co-chair of the industry-wide Quant UX Con (`www.quantuxcon.org`). His work also engages closely with quantitative marketing research, where he has chaired three analytics conferences and coauthored two popular texts, *R for Marketing Research and Analytics* [25] and *Python for Marketing Research and Analytics* [127]. These have been used as textbooks at more than 100 universities.

Kerry wrote the first set of criteria for hiring Quant UXRs at Google. Chris later led teams that updated and expanded the Quant UXR interview and hiring process, and served on the overall Hiring Committee for all UXRs. Over the years, Kerry and Chris have interviewed more than 200 Quant UXR candidates and mentored more than 40 Quant UXR employees.

We hope our shared knowledge will help you to accelerate your career!

1.2 What Is Different About This Book?

There are many books that describe the technical skills used by Quant UXRs, including texts on research design, programming, statistics, and user research. However, there is no other book that goes deep into the *role* and *practice* of Quant UX research, describing what is unique about it and how Quant UX relates to other aspects of research and development.

We will not try to teach all of the technical skills that are needed for Quant UXR. Many books covering those skills exist and we reference them so that you will be able to find what you need to know. (The complete list of references appears in Appendix C.) What we add is context: among all the things you *could* know, what do you *need* to know? What are examples of Quant UXR work? How is the role different from other roles? What will help you succeed, going beyond mere technical knowledge?

In short, our goal is to share things that you would otherwise only learn after many years *on the job*, not in a textbook.

Along the way, we are explicitly and occasionally strongly opinionated. There are trends in Quant UX that worry us, along with practices that we believe are mistakes. We encourage you to view our opinions as part of a friendly debate. Although we hope to be convincing, our goal is not to change your opinion. Rather, our opinions should provoke you to think about issues. If you disagree with us while learning something new or strengthening your convictions, that is a great outcome.

1.3 Who Is Our Audience?

We are writing for anyone who wants to learn about Quant UX and gather advice and guidance from longtime practitioners. In particular, we address the following readers:

- Quant UXRs who are looking to accelerate their careers

- General UXRs who want to understand more about Quant UX research

- Data scientists who are involved in user research or working with UX partners

- UX managers who are hiring or managing Quant UXRs

- Anyone considering switching into the field from academia or other roles

- Anyone interviewing for a role as a Quant UXR

As we previously noted, this book is not a textbook about the technical basics of Quant UX. We assume that you have or will acquire the disparate technical skills of the job, and we describe those skills and the level of expertise needed. Our job here is to help you to put the skills together, apply them to projects, use them better, and succeed as a Quant UXR. It might instead convince you *not* to become a Quant UXR, and instead to look to data science or another career—and that is also a valuable outcome. Either way, the methods and approaches here will help you understand users.

1.3.1 A Quick Check on Your Interests

How would you answer the following questions?

1. Do you understand basic descriptive and inferential statistics? For example, could you explain to a non-specialist exactly what a *standard error* means, or what a *t-test* does?

2. Do you prefer to use a language environment—such as R, Python, or MATLAB—to do statistics, instead of a point-and-click or scripted tool (such as JMP, SPSS, or Excel)?

3. Can you program at the level of writing reusable functions?

4. Do you *enjoy* programming?

5. Do you enjoy tackling new problems regularly, while solving them quickly and approximately?

6. Do you have experience (or at least a strong interest) in human behavioral research and research design?

7. Are you motivated to make products better for users' needs?

If the answer to most of those is "Yes!" then a Quant UXR role might be a good fit for you. We will say much more about these topics, and how the role differs from other roles. In Section 3.5 we present a brief quiz to compare your skills for Quant UXR and several other roles.

1.4 What Will You Learn?

This book goes beyond textbooks, cookbooks, and reference guides to discuss things that would be difficult to learn without years of experience. Those topics include

- What really *is* the role of a Quant UX researcher?

- How does Quant UX research differ from other areas such as data science?

- How does the Quant UXR role fit into an engineering or UX organization?

- What are typical UX projects, and how do they run into trouble?

- What does it take to succeed as a Quant UXR? What is necessary to know, and what is unnecessary?

- What should you expect in job interviews?

- What does career progression look like?

We share the same advice we would give colleagues or executives when we are chatting, mentoring, advising, hiring, or debriefing performance reviews.

At the same time, we cover important technical topics that we believe are not adequately covered in existing texts. These include *concepts* that we find useful across many projects, conceptual *frameworks* that can help you to plan and communicate research, and *methods* that may be new to you and could expand your skills. The technical areas are

- The HEART model (happiness, engagement, adoption, retention, task success) to help define metrics of user experience using a memorable framework

- Assessing customer satisfaction (CSat), where we address both survey design and analytic methods

- Visualization of sequences in logs of product usage data, applying a "sunburst" visualization

- MaxDiff survey design and statistical analysis, which helps prioritize user needs, features, and messages

How about *code*? When it is relevant—as in discussions of programming and specific statistical methods—we include R code that you may run for yourself or reuse in projects, and we provide example data sets.

1.5 How to Use This Book

We have designed the book to be read either straight through or by skipping around. The chapters follow a logical progression (we hope), yet each one may be read by itself, in any order.

Here are recommended starting points:

- *If you are new to Quant UX*: Chapter 2, "User Experience and UX Research," Chapter 3, "Quantitative UX Research: Overview," and Chapter 11, "UX Organizations," plus chapters on selected skills.

- *If you are a current Quant UXR*: Chapter 3, "Quantitative UX Research: Overview," Chapter 4, "UX Research," and Chapter 14, "Career Development for Quant UX Researchers." Although we truly hope you'll find all of the book interesting!

- *If you manage Quant UXRs*: Chapter 3, "Quantitative UX Research: Overview," Chapter 11, "UX Organizations," Chapter 13, "Research Processes, Reporting, and Stakeholders," and Chapter 14, "Career Development for Quant UX Researchers."

- *If you are interested in example projects and approaches, rather than the role of Quant UXR*: Chapter 3, "Quantitative UX Research: Overview," Chapter 7, "Metrics of User Experience," Chapter 8, "Customer Satisfaction Surveys," Chapter 9, "Log Sequence Visualization," and Chapter 10, "MaxDiff: Prioritizing Features and User Needs."

Each chapter ends with a summary of key points and a list of recommended sources to learn more. Fair warning: we love books, and our references might convince you to acquire a few more for yourself!

When we cite a book, article, or other material, we use bracketed numbers that correspond to entries in Appendix C. For example, if we refer to "a classic C text [68]," the corresponding numbered entry in Appendix C is Kernighan and Ritchie, *The C Programming Language*.

1.5.1 Assumptions

It is impossible to cover every aspect of every topic or to define all the variations of every role, method, and approach. To simplify the task, we make the following assumptions and choices:

- We write about roles and projects in the *tech industry* and organizations that have similarly constituted product engineering teams. Our research topics apply to product design and engineering for computer software, hardware, applications, and services. However, the approaches and methods might be used far more widely, as in human services, public policy, or many other areas.

- Our focus is on the *role* of Quant UX researcher and on the methods that Quant UXRs use and that might be used by others. When we discuss job roles, as opposed to methods, we mostly refer to full-time, permanent employee positions.

- We discuss research with *people*. We assume that the data of interest are generated by human actions, even if indirectly, and that the research participants are people.

- We mostly use *Google UX terminology*. Job titles and other terms vary from company to company and we can't list every variation. We adopt Google terminology as a standard because we are familiar with it, it is where Quant UX started, and it has the most Quant UXRs as of the time of writing in 2023. Although smaller organizations may not have all of the kinds of data, resources, and roles that we discuss, the concepts and themes apply to organizations of any size.

- We describe the *most common* variations of topics, roles, and methods. Any role—research, design, engineer, manager, etc.— does many different things, and we don't detail everything that someone might do. Instead, we present the core activities as we understand them.

Most of our topics and discussion apply to any field or organization outside of "tech" that shares interest in quantitative understanding of behavior; this could include medicine, architecture, aviation, the military, hospitality, and entertainment, among others. In some cases, another field may look almost identical to the tech industry with regard to product development. In other cases, there may be substantial variation. Use your own domain expertise to draw connections between our tech examples and possible application in other fields.

If you are interested in a specific company, you'll want to translate our definitions into theirs. For example, an organization may have "design research" or "user research" instead of "UX research," or "developers" instead of "software engineers."

1.5.2 A Note About Jargon

We generally try to define terms and acronyms when we first use them, and occasionally thereafter. We will often put those in *italic* when first introduced. However, our assumptions about what should be defined might not match your expectation. Your friendly search engine is a good resource.

We use industry jargon, UX slang, and abbreviations for the purposes of this book, sometimes without defining them. That is intentional. It allows more concise communication and, even more importantly, is part of your acculturation as a *UXer* (see how that works, when we slip in a new term?) As previously noted, we use Google circa 2023 as the standard source of terminology.

When you understand and use the jargon in the field, you will be that much closer to doing well in interviews, communicating with colleagues, and building on the concepts to develop more complex ideas.

1.5.3 End of Chapter Exercises

Where we believe it is relevant—especially in technical chapters with code—we include exercises for you. Some are thought exercises asking how you would explain a concept or apply a method to new data sources. Other exercises ask you to write code to solve a particular problem. Those exercises use data from the chapter, or that you generate, or that is hosted online.

The coding exercises are designed to do three things: to reinforce your understanding of the concepts in the book, to build your skills in writing code to work with simple yet realistic data, and to demonstrate the approximate level of coding skill that we *recommend* for Quant UX researchers (even if it is not strictly required; see Chapter 6). The exercises are *not* intended to be examples of possible interview questions; they are often more specific or more difficult.

We strongly encourage you to complete the exercises! Although we don't provide solution code for them (because we want *you* to complete them), we can say that every one of them was solved by us in a moderate amount of time. We hope they are helpful and enjoyable for you, and our main recommendation is not to treat the exercises as *tests*, but rather as learning opportunities.

1.6 Online Materials

The companion website for this book is `https://quantuxbook.com`. At the website you can find R code and example data sets that are referenced in the technical chapters and example analyses. In each chapter where we present code, we give a URL for the .R code file, and each .R code file will load its own data from another URL that is either a CSV (comma-separate value) or an RDS (native R) data file.

If you will be working offline, or need to access data that may be blocked by your network, you may instead download the code and data files in advance. Please see the website to find a ZIP file to download all code and data files.

1.6.1 Code and Data Sources

All code and data in this book are either newly created here or have been previously published. None of the code or data here comes from the current or previous employers of either author.

Specifically, none of the data was collected from users or customers of Amazon, Code for America, Google, or Microsoft, and none of the code or data is proprietary to any of those firms. We do use some open source code to which they contributed.

You are welcome to reuse any of the code in your own work in accordance with the MIT License (`https://opensource.org/licenses/MIT`). A citation of this book would be appreciated but is not required. The code here was written for the educational purposes of this book and its suitability for any other project is up to the user to determine. As noted in the license, "The software is provided 'as is', without warranty of any kind, express or implied."

1.6.2 Help! Updates and Errata

We have tried to make the book error-free. All of the technical details were checked immediately before publication. However, things change. If you run into technical problems, try these suggestions:

1. If you disagree with us, that's no problem! Our goal is to provoke thought, not to convince you of every opinion.

2. Check the errata page on the book's website `https://quantuxbook.com` to see whether we have addressed the problem.

3. Our code files run in R, and the book was completed using R 4.2.2 running in RStudio 2022.12.0.353 on Mac OS X 13.1, with all packages updated on January 30, 2023. Small details in statistical models may depend on versions, OS, or random number libraries. If the result is *close* but not exact, you're probably OK.

4. If something crashes, try turning it off and back on again. In other words, reboot! (This is also a good time to enjoy an episode of a great TV show that recommends that solution, *The IT Crowd*, `www.channel4.com/programmes/the-it-crowd`.)

5. If all else fails, email the authors. We can't answer every question, but we promise to read them and will help when possible. Email: `quantuxbook@gmail.com`.

1.7 Key Points

If you're looking to understand the role of Quant UXR from the inside out—what it is, what you need to know, and how to excel using quant UX methods or working as a Quant UX researcher—you're in the right place.

If you're looking for a textbook that will tell you exactly what to do, such as how to program or do general statistics…this is not it. We cover a few technical projects but those assume basic knowledge of statistics and programming. You may be more interested in *R for Marketing Research and Analytics* [25] or *Python for Marketing Research and Analytics* [127]. We refer to those texts often as the R and Python "companions" to this book. We even use some of the same data sets in order to build on the skills taught in those books.

To start learning more about Quant UX, turn the page and let's begin!

CHAPTER 2

User Experience and UX Research

In this chapter, we answer the questions, "What is UX?", "Who works in UX?", "What is UX Research?", "How does UX relate to product development?", and "What are typical UX research activities?"

This chapter will be of greatest interest to readers who are not familiar with user experience (UX) organizations. We briefly describe the common roles in a UX organization, present typical UX research tasks, and describe how they relate to the *lifecycle* model of product development. By the end of this chapter, you will understand the general structure of how UX organizations function and what UX researchers do.

Readers who *are* familiar with UX organizations may wish to skim this chapter before reading about quantitative UX research in Chapter 3, "Quantitative UX Research: Overview." In that case, we want you to take away one key point: we strongly believe that quantitative UX research is a subspecialty that falls within UX research.

2.1 User Experience

User experience (UX) encompasses how a product's users perceive, learn to use, understand, interact with, accomplish tasks with, and generally think about a product.

A *product* in this sense is anything designed for human usage or interaction. Most commonly, the term "UX" is used for technology products, such as computer software, hardware, and consumer electronics devices. Such products are made not only by traditional technology industry companies but also by firms in many other fields including finance, healthcare, retail, hospitality, manufacturing, transportation, and governmental agencies. Banks make user applications, government agencies have

© Chris Chapman and Kerry Rodden 2023
C. Chapman and K. Rodden, *Quantitative User Experience Research*,
https://doi.org/10.1007/978-1-4842-9268-6_2

websites, amusement parks design visitor experiences, and manufacturing companies develop internal control applications. Any of these organizations could have a team devoted to understanding and improving UX of their products.

However, UX in the broadest sense exists everywhere. A door has a user experience [101], as do houses, parking lots, books, clothing, hiking trails, food, chairs, faucets, churches, doctors' offices, prescription medicine, paint, automobiles, maps, space shuttles, and every other human-made good, service, and environment. All of them involve learning, interaction, and potential success and frustration.

Although we hope you will keep that breadth in mind, in this book we'll apply the term "UX" primarily to computer software and hardware, consumer electronics, and similar systems. Why? To put it simply, those are the areas where most UX practitioners work—whether in a technology company or on a technology and UX team that is part of a company in a different industry. As far as we know, all Quant UX researchers currently work on such products.

On the other hand, the skills and approaches here apply widely and they may be used across the full range of products and services just noted.

2.1.1 UX Roles

In general terms, user experience teams are responsible for the design of the parts of a product that users interact with. Software engineers—again, in general terms—are responsible for writing the code to implement that design, so that it is efficient, scalable, maintainable, and reliable.

A UX team may include several of the following roles, among others, although the exact titles may vary widely:

- UX designer

- UX researcher

- UX manager

- UX writer

- UX producer

- UX research coordinator

The role of the *UX designer* (UXD) is to create the interactive experience of the product. A UXD may specialize in interaction design, industrial design, visual design, or other aspects of the experience (there are specialists in video design, sound, typography, and many other areas). In this book we refer simply to *designers*. Among them, researchers most often work with specialist interaction designers and visual designers.

UX researchers (UXRs) engage in qualitative and quantitative research to answer questions about users and their interaction with, or their perceptions of, a product. These questions may range from purely descriptive (for example, "What are users doing?") to inferential (such as, "Which product would they prefer?"), and may include both primary and secondary research.

A *UX manager* leads a UX team, providing guidance and oversight of individual team members. The manager engages with peer teams, managers of other disciplines such as engineering, and executives. UX managers may have a background in any UX area, and may manage team members with different backgrounds. For instance, it is not unusual for a UX researcher to have a UX designer as a manager. We discuss UX management and team organizations in Chapter 11, "UX Organizations."

UX writers oversee the linguistic, textual, and similar content of a user interface. They are responsible for the content being brief, clear, accurate, and generally understandable and helpful to the user. Besides writing text, labels, and so forth, a UX writer works with UX designers to establish the *voice* of a product, such as its degree of formality or informality. A UX writer works to ensure consistency of language and voice across the product, and in supplementary contexts such as help pages. Note that this is quite separate from *marketing communication* or *technical writing*; UX writing is about the design of language within the product itself.

Some teams have the role of *UX producer* (sometimes called a UX program manager or design producer). A UX producer coordinates multiple aspects of the UX design and research process. For example, a UX producer may coordinate the overall activity of multiple designers and user researchers working on different parts of a complex product. This involves strong skills in project management and communication as well as design.

Finally, a *UX research operations* (ResearchOps) specialist is a project manager who supports a UX research team. This may involve overseeing how users are recruited for research studies, managing research lab spaces, handling data from studies, working with external research facilities, and everything else that goes into making research run efficiently. In large companies, this role is often centralized, with a single team supporting multiple UX groups across multiple products. However, when possible, it is advantageous to have a dedicated ResearchOps specialist for each UX team, because they will develop better understanding of a product's user population and effective tactics to recruit those unique study participants.

2.1.2 UX Design and Software Engineering

How does UX relate to software engineering? Let's consider an e-commerce website. UX designers would plan the layout of the pages, how information is presented to users, and the series of steps that would be needed to complete tasks, such as the buttons, menus, shopping cart, and checkout process. Designers also specify the fonts, graphics, color scheme, how the site responds on mobile devices, how users obtain help, and so forth.

Software engineers (SWEs) implement most or all of the code that delivers that user experience. This includes everything from *back-end systems* that provide databases of users, products, inventory, and product reviews; *middleware* that connects the back-end systems to the front end, and also may engage with partner services such as cloud services that scale up to meet sudden demand or billing services to process orders; and *front-end programming* of the user interface itself, implementing the visual design, navigation structure, handling for various browsers and devices, and so forth.

Such an ideal distinction between UX and software engineering rarely exists, and the roles often overlap. For example, there are typically many more SWEs than designers, and engineers will develop some aspects of a product's user interface on their own. Designers may be brought in for only the most "important" parts of the product. On the other side, designers may write code to implement their designs, rather than delegating that to a software engineer. When designing a user interface, it is often a good idea to create a high-fidelity *prototype* in code. That allows designers to demonstrate the exact interaction model and researchers to conduct more realistic user testing. Later, the prototype might be discarded, although in some cases parts of the code could end up as part of the final product (adding much more code for security and robustness).

We would highlight SWEs who work in *Test* (SWETs). They are specialized SWEs dedicated to code and product reliability. SWETs work to find and eliminate bugs, check and improve code quality, identify and mitigate security threats such as vulnerable user data, and ensure reliable, fast performance across devices and real-world conditions. SWETs often enjoy observing UX research studies, because direct observation of users is a great way to find bugs.

2.1.3 Product Management

A product team includes other important roles. For many UXers, a crucial partner is the *product manager* (PM; sometimes called a "program manager," although that may also be a different role). A PM serves as the connector between the larger business and the engineering team and is responsible for guiding the engineering team to deliver a successful product.

PMs inform and guide decisions about product capabilities and priorities. They help determine the priority of features to implement and how the importance of a given feature compares to the available engineering resources and schedule. PMs work with UX and marketing to define the target customer for the product, and just as importantly, who is *not* the target customer. A key responsibility of PMs is writing a product requirement document (PRD) that defines what the product will do. The PRD details the important features for the product, specifies its expected price for customers, and may include information such as a target schedule for development.

Although a PM is called a "product manager," that does not imply that a single PM is also responsible for every aspect of what customers perceive to be a "product." Instead, PMs are usually responsible for a few parts of a larger product. Consider a fictional word processing application, which we'll call Zenith Write. A customer may perceive the entire application Zenith Write to be the product. However, from a software engineering point of view, Zenith Write comprises many distinct parts, such as an editor window, a spell checker, printing capabilities, a drawing tool, and so forth. These components may have different PMs, and each component, from the perspective of PMs, could be regarded as somewhat independent. Each PM leads decisions for their scope of the product, while PM managers guide decisions of larger scope, such as the entire product, Zenith Write.

The most important thing about the PM role is this: the PM is not the boss but rather is an influencer, adviser, and (sometimes) decision maker who works closely with other team members. A PM learns about potential product capabilities and limitations from SWEs; about user needs from UX researchers; about visual and interactive design options from UX designers; about market trends and needs from marketing research; about corporate strategy from upper management; and so forth. The PM's job is to bring together all of those sources of information and coordinate how decisions are made about the product.

You may be thinking, "A PM sounds like the perfect person to influence with UX research." We agree! PMs are often the closest colleagues for UX researchers.

2.2 UX Research

User experience researchers (UXRs) conduct primary (new) research studies, along with secondary reviews of existing research, to answer crucial questions about what users need, how they use a product, how satisfied they are, what might be improved, and other aspects of their experience with a product.

In the early days of user research, UXRs were known as *usability engineers*, with the assumption that their primary job was to make a product more *usable*. The research often involved observation of users' problems in lab-based *usability tests*. Usability pioneer Jakob Nielsen outlined five key areas to consider [99]:

- *Learnability*: How easy is it for users to start using a product fruitfully?

- *Efficiency*: Can users accomplish their tasks in an efficient way?

- *Memorability*: Will users retain their learning and continue to be efficient in later usage?

- *Errors*: Does the product operate without error, and do users commit few errors?

- *Satisfaction*: Are users happy with the product?

This is a classic human-computer interaction (HCI) model that presumes that user behavior may be studied in terms of users' *goals* and the *tasks* performed to meet those goals. It measures interaction efficiency such as the time taken to complete a task, or the number of keystrokes [13] and the errors and misunderstandings that arise during the tasks.

In UX today, these considerations remain important for many products. However, the role of UX research has broadened considerably to include many other topics beyond those that can be studied in a usability lab, such as exploratory field research to understand user tasks in context, definition of target customers, and analysis of actual usage patterns. Nearly every industry has seen widespread adoption of interactive technologies that are crucial to an organization's operations and success. This is leading to the dispersion of UX researchers and research activities across increasingly diverse areas.

2.2.1 Categories of UX Researchers

The most common category of UX researcher is simply a *UX researcher* without a more specific designation. In this book, we refer to this as *general UX researcher*. We use the term *UX researcher* to refer to a UXR of any category (general, quant, or other).

The distinction of general UX researcher vs. quantitative UX researcher is often made by referring to general UXRs as *qualitative UX researchers*. This follows the notion that if they are not quants, they must be *qualitative* researchers and that their research often involves small sample sizes. We find the *qualitative* title to be misleading, because many general UXRs use quantitative methods and may not have in-depth training in traditional qualitative methods such as ethnography. To solve that, the term *mixed-methods research* is sometimes used, although that is even more confusing. What does it mean to mix methods? Does anyone use only one method? We avoid these questions by using the term "general UXR."

Similarly, in some organizations, a general UX researcher has the title *usability engineer* or *design researcher*. Those titles, unfortunately, suggest that UX research exists only to assess traditional usability or the work product of designers.

Because this book is about quantitative UX research, we distinguish Quant UXRs from general UXRs and ignore other research titles and distinctions. However, there are a few specialties worth knowing about, as follows. We list them as job titles, although a researcher might have any title while performing the functions noted.

- *Ethnographer*: Conducts in-depth field research to understand users and their needs in real-world contexts.

- *Usability researcher*: Assesses product designs for usability and provides detailed recommendations for design changes.

- *Human factors engineer (or ergonomics engineer)*: Studies the physical engagement of a user with a system, such as touch, mechanical usage, vision, perception, and so forth. Classic examples include research in aviation, military systems, nuclear power, automotives, and similar environments where high cognitive demand is paired with high risk.

- *Survey scientist*: Designs, fields, and analyzes surveys to assess user needs and product experience.

A team of UXRs may staff these positions according to the product need. For example, a hardware product such as a keyboard or smartphone may need a human factors engineer, whereas a product that a company wants to expand to new markets may benefit greatly from ethnographic research. Individual UXRs often develop experience in one or more of those areas.

2.2.2 The Research Lifecycle for UXRs

UX research usually follows the product development *lifecycle* shown in Figure 2-1. In practice, a researcher might enter the cycle at any point, but it is most natural to think about creating a new product from scratch, in which case they would start at the top. Beginning at that point in Figure 2-1, a product is inspired by an unmet need in the market and progresses through initial development, refinement, and product release.

Each stage in the lifecycle has typical UX research activities. In early stages, UX research may be heavily engaged in identifying crucial product needs. As development progresses, research engages more with assessing the product's performance and how well it meets users' needs. After a product is released, UXRs may identify new behaviors or unexpected issues, which then lead into the needs for the next version or future products.

There are three important things to note about the lifecycle. First, the stages overlap in practice, because different portions of a product will be in different stages of development at any given time. Consider our earlier example of a word processing application. The editor may be very mature and remain essentially unchanged from version to version, and be subjected primarily to usability testing and assessment of satisfaction. At the same time, a new feature such as an AI-based capability may require early-stage needs assessment. UXRs may work across multiple stages, or specialize in one or two stages, such as usability lab assessment or survey research. We describe specific research activities in upcoming Section 2.2.3.

The sequential layout of Figure 2-1 is an idealized model that does not exist in reality. In particular, long-range planning should occur throughout all stages because user needs are dynamic, and teams must respond to competing products; some features may be postponed to later releases.

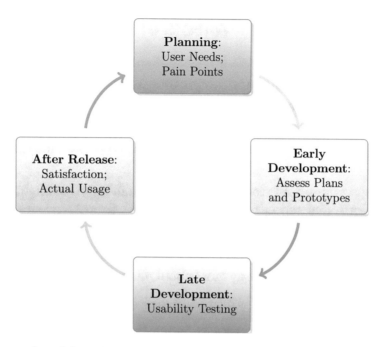

Figure 2-1. *Product lifecycle and typical UX research activities. UX research may begin to assess users' needs before engineering starts (for a given product release), evaluate the product in development during the engineering phases, and continue to understand the real-world experience after the product launch*

Second, some UX research activities are not tied to the lifecycle model. *Foundational* research establishes knowledge about users and their needs that informs all stages of product development. For example, a user segmentation model (dividing users into different types) may inform a product holistically. Or, for medical equipment, airplane cockpits, keyboards, and many other physical products, anthropometry (study and measurement of human bodies) may be an important and ongoing research topic separate from any particular product plan. UX research may also undertake *generative* research, whose aim is to uncover user needs that suggest new products or areas for innovation.

Third, the research needs in any given stage should include a mixture of qualitative and quantitative research. Research findings are more complete and influential when they combine insights into both "why?" (more qualitative) and "how much?" (more quantitative) aspects of users' experiences. This is one reason we believe UX research is a unified discipline even when researchers specialize in various methods.

2.2.3 Typical Research Projects in the Product Lifecycle

Each stage of the product lifecycle (Figure 2-1) presents specific questions for UX research. Generally speaking, there is a progression from relatively broad questions in the planning stages to narrower, more specific questions as development progresses. After product release, the questions broaden again as the team moves into planning. As we previously noted, this is a schematic simplification; in practice, the stages overlap and intermix.

Table 2-1 lists common UX research projects and methods that might be applied to each stage of the lifecycle. For our purposes, we will not detail each of them, because they are mostly addressed well in the broad UX research literature in terms of *general* UX research. For pointers to those discussions, see the final section of this chapter, "Learning More" (Section 2.4). We say more about many of these projects from a *quantitative* perspective in later chapters.

Table 2-1. *Typical UX Research Projects Across the Product Lifecycle*

Stage	Typical Research Projects
Planning	• Ethnography, interviews, customer site visits • Task and workflow analysis • Identify pain points and blocking issues in tasks • Feature preference and product needs assessment • Assessment of competing products • Customer segmentation
Early Development	• Testing with interactive prototypes • Feature trade-offs, willingness to pay • Competitive preference
Late Development	• Usability testing • Early adopter "beta" testing • A/B experiments for feature alternatives
After Release	• Satisfaction survey • Logs analysis of usage and related behavior • Diary study of product usage • Experience sampling

No UXR can be an expert in every kind of project shown in Table 2-1. A reasonable expectation is that a UXR will develop a *T-shaped* set of skills. That means that a UXR will demonstrate basic competence and breadth (represented by the head of a *T*) across many methods while showing expertise (the descending tail of the *T*) in a few. We say more about T-shaped skills in Chapter 4 (Section 4.1.1).

Expectations for a UXR (and the coverage of their "T") changes over a career. In the early stages of a career, a UXR might specialize in only one area, according to their educational or other background, such as usability assessment, survey analysis, or ethnography. As UXRs develop expertise they tackle new projects and questions that concern a larger scope across a product. In the senior levels of a career, a UXR is expected to demonstrate high competence in many—perhaps half, or more—of the projects shown in Table 2-1, with deep expertise in a few of them.

2.3 Key Points

From this chapter, you should be familiar with the basic concepts of user experience and UX roles, and how those relate to UX research. Important points are as follows:

- UX is part of product development and exists alongside software engineering and product management (Sections 2.1 and 2.1.2).

- UX roles include design and user research, along with other positions (Section 2.1.1).

- A simplified way to consider product development is with a circular lifecycle model that begins with product planning, moves through engineering, and leads to product release. In reality, these stages tend to overlap, but the model is conceptually useful (Section 2.2.2).

- UX research aligns in large part with the product lifecycle. Research activities transition from early assessment of user needs to narrower, focused product evaluation as the cycle progresses (Section 2.2.3).

- In addition to research assessing specific product designs and decisions, UX research may also engage in foundational research that informs a product and business strategy independently of the lifecycle (Section 2.2.2).

- There is no clear distinction in UX research projects between strictly qualitative or quantitative research. Most projects could be addressed with either a qualitative approach or a quantitative approach, or a mix of the two (Section 2.2.2).

- Individual UX researchers have personal expertise and interest in a subset of possible UX projects and methods. Over the course of a career, a UXR should develop additional breadth and depth, sometimes known as a *T* shape (Section 2.2.3).

2.4 Learning More

Don Norman explains how to think like a user experience professional in a book written for non-specialists, *The Design of Everyday Things* [101]. The well-chosen (albeit sometimes dated) examples will inspire you to identify user experience problems almost everywhere. Although the content is non-technical, for those who are new to UX, it is well worthwhile to develop the kind of intuition that Norman demonstrates (and conversely, if you don't enjoy such analysis, UX might not be the right field for you).

An overview of UX organizations, from the perspective of UX researchers, is given by Baxter, Courage, and Caine in *Understanding Your Users: A Practical Guide to User Research Methods* [4]. After describing the role of UX research, the authors describe general UX research activities in detail. The discussion includes practical considerations such as participant recruiting and lab management, along with instruction on how to conduct common user research activities such as diary studies, interviews, and product evaluation sessions. It is an excellent introduction to the current practice of general (sometimes called "qualitative") UX research.

For evaluative studies aligned with the "late development" stage (Section 2.2.2)—and especially those in a *usability lab*—a foundational and memorable text is Nielsen's *Usability Engineering* [99]. Nielsen outlines crucial areas to assess and discusses important topics such as the cost/benefit trade-off among methods, sample size requirements, how to conduct laboratory evaluations, and *discount usability* and *heuristic assessment* alternatives to traditional studies.

CHAPTER 3

Quantitative UX Research: Overview

What is Quant UX research? And, just as importantly, what is *not* Quant UX research? If you are wondering whether it might be a good fit for you, this chapter will help you answer those questions.

First we look at what Quant UXRs do and the kinds of problems they tackle. Later in the chapter we discuss how the Quant UXR role differs from other roles such as data scientist and general UX researcher.

3.1 Quantitative UX Research

We'll start with our definition: *quantitative UX research is the application of empirical research methods to inform user-centered product design at scale.*

Let's break that down, starting from the end. *At scale* means that Quant UXRs are able to consider projects with any appropriate amount of data. In some cases, an appropriate amount may be billions of observations from a product log or website; in other cases, it may be only a few sets of observations from a small number of crucial users. Quant UXRs know when to use massive data sets, and when it is appropriate to use a random sample from a data set. On the technical side, Quant UXRs have skills in statistical reasoning, programming, and database management to identify the right scale and execute projects as needed. (See Chapter 5, "Statistics," and Sections 6.1 and 6.4.2.)

Product design means that Quant UXRs are concerned primarily with addressing questions that arise in design and engineering. The role has little or no focus on finance, demand forecasting, business operations, sales, channel marketing, or many other areas related to business data. Those areas are often the focus of data scientists and business analysts, as we discuss later in Sections 3.4.5 and 3.4.6. Those are important areas yet they are tangential to product design.

© Chris Chapman and Kerry Rodden 2023
C. Chapman and K. Rodden, *Quantitative User Experience Research*,
https://doi.org/10.1007/978-1-4842-9268-6_3

User-centered implies that Quant UXRs are primarily focused on *people* and the kinds of behavioral and attitudinal data that people generate. This implies two things: first, it somewhat downplays data created indirectly by people, such as data from logistics, finance, data center operations, machine, device, and software performance; second, it is not primarily concerned with data generated by businesses, processes, automation, devices, and the like. User-centered research asks, what are *users* doing and why? (See Section 4.2.)

Application of … research methods defines that Quant UX research is a *research* activity, meaning that it is concerned with research design along with primary data collection and analysis. Quant UXRs should be involved closely with research planning and the design of data collection. They are responsible for choosing appropriate analytic methods, interpreting analyses, and communicating the results to others. Quant UXRs do not simply obtain and analyze data from others.

Empirical research is an imperfect way to say that the role is concerned with observational or experimental data gathered from the world. That contrasts with work that engages primarily with theory, logic, data simulation, or programming.

Each one of those aspects of *quantitative UX research* has fuzzy boundaries, and Quant UXR is imperfectly differentiated from other roles. Some general UXRs and data scientists, among others, may engage in similar research even when their overall role is somewhat different. In this chapter, we attempt to clarify the boundaries and advocate how we believe Quant UX research *should* be constituted.

3.2 Week-to-Week Practice of Quant UX Research

Quant UXR work doesn't align with *daily* tasks in any particularly useful way. An individual day may be dominated by meetings, coding, travel, or training. Instead, it is better to consider what Quant UXRs tend to do over the course of a *week*. We find that meaningful projects typically take at least a week; and units of weeks may be useful when estimating the effort required for projects.

By the way, we do not use *week* to mean 40 hours of *effort*, but rather use it as an amount of elapsed time. What's the difference? A project might require an extended duration even though the effort on each day is small. It may take weeks to complete data collection for a survey, during which time one is free to work on other projects.

3.2.1 Typical Activities in a Week

The most important activity of a Quant UXR in the course of a week is to meet with, check in, or otherwise interact with the key stakeholders for your research. A *stakeholder* is anyone who has particular interest in the outcomes and results of your research, especially when they are charged with making decisions about a product (or your career). Ideally, the most important stakeholders will be engaged from the beginning of a project to help define the research question, describe what they would do with the results of research, and refine the kind of answer that would be useful for their decisions. Chapter 13, "Research Processes, Reporting, and Stakeholders," discusses relationships with stakeholders.

Typical stakeholders for Quant UXR projects are product managers (PMs), UX designers, software engineers (SWEs), executives, and, importantly, UX research colleagues. See Chapter 2, "User Experience and UX Research," for descriptions of those roles.

The most crucial form of engaging with stakeholders is presenting research results (typically a slide deck or other report, presented live). However, despite the high stakes of such presentations, they are also comparatively less common than other ways of engaging with stakeholders; a Quant UXR should not work in obscurity and only appear to give a presentation. Rather, research results should come as the natural culmination of a longer path of engagement. When the stakeholder is engaged along the way, the research will be better targeted and more influential (see Chapter 13).

After engaging with stakeholders, the second most important set of tasks in a given week is to advance one or more (usually several) projects. The activities depend on the research lifecycle phase for your Quant UXR projects (see Section 2.2.2). You may give attention to defining research problems, collecting data, writing code, building analytic models, and writing reports or presentations of the results. Some weeks are all about data collection while others are devoted to writing analytic code. But over the course of a few weeks, a researcher will engage in many activities.

A third set of tasks is organizational housekeeping. This includes all the things that make an organization run: staff meetings, one-on-one meetings with your manager, training, travel, expense reports, performance reviews, interviewing candidates, mentoring, and so forth.

An unfortunate reality is that the time taken by each set of the activities just described is sometimes *inversely related* to research. It is not unusual to spend the largest proportion of your time on the third set of tasks—organizational housekeeping— and less than you would like engaging with analyses, writing code, and meeting with stakeholders. Strong management may help to rectify that imbalance; we have suggestions in Chapter 11, "UX Organizations."

3.2.2 Common Research Questions for Quant UXRs

Referring to the research lifecycle model in Section 2.2.2, Quant UX research questions vary according to the product stage. In Table 3-1, we repeat the general UX research table from Chapter 2 but retain only the projects that commonly involve Quant UXRs.

Many of these projects commonly also involve paired qualitative research, and we mark those with an asterisk in the table. For example, in a customer segmentation project, a Quant UXR may work to identify clusters in the data using statistical methods while a general UXR conducts interviews with users to build descriptive profiles of the segments.

Table 3-1. *Quant UX Research in the Product Lifecycle*

Stage	Research Activity (* involves qualitative)	Typical Quant Methods
Planning	• Task and workflow analysis* • Feature preference and product needs assessment • Assessment of competing products* • Customer segmentation*	Logs analysis MaxDiff choice modeling Conjoint analysis, brand perception Clustering, classification Conjoint analysis
Early Development	• Feature trade-offs, willingness to pay • Competitive preference*	Conjoint analysis A/B and multivariate experiments
Late Development	• A/B experiments for feature alternatives	In-product or online survey Logs analysis
After Release	• Satisfaction survey* • Logs analysis of usage and related behavior • Experience sampling* • Define and measure outcome metrics • Feature impact on an outcome metric	Various, such as ecological momentary assessment Construct definition and assessment Causal modeling

3.2.3 Stakeholder Questions

Of course, a researcher does not start with the method first, but instead designs the project according to stakeholder questions. In Table 3-2, we outline typical questions that stakeholders ask, paired with common quant UX projects to address them. You can trace a question in Table 3-2 to the related methods in Table 3-1.

Table 3-2. *Stakeholder Questions in the Quant UX Lifecycle*

Question	Project (* also involves qualitative)
"What do users do with our product today?"	Task and workflow analysis*
"Which features should we prioritize for the next version?"	Feature preference and needs assessment
"How do we stack up against our competition?"	Assessment of competing products*
"We want to understand our user types more deeply."	Customer segmentation*
"How much will users pay for feature X? For our product?"	Feature trade-offs, willingness to pay
"How many customers will buy our product vs. the product of competitors?"	Competitive preference*
"Does design A or design B drive higher usage/success/etc.?"	A/B experiments for feature alternatives
"Are our users satisfied? What should we improve?"	Customer satisfaction survey*
"Are we hitting our goals for user adoption and retention?"	Logs analysis of usage behavior
"How are users actually using the product, moment to moment?"	Experience sampling*
"Does feature A drive higher usage/success/etc.?"	Feature impact on an outcome metric

There are important points to note about the questions and projects in Tables 3-1 and 3-2. First, the particular research methods in those tables are not the only ways to address specific stakeholder questions. Different Quant UXRs have experience with different methods and should select approaches that are well suited to their expertise and interests.

Second, not every stakeholder question is a good question or worth the effort to answer. There is *opportunity cost* in research, and any particular project that is done means that some other project will not be done. It is important for Quant UXRs to work with stakeholders to select appropriate projects. This is a skill that develops with experience, as you gain insight into the usual course, timing, effort, and pitfalls of various methods. Later in this book we share our experiences that may help you to address those trade-offs. Chapter 13 discusses common problems with stakeholder expectations.

Third, not every Quant UXR can be an expert in every method. In the next section, we describe common patterns among Quant UXRs in skills, background, and expertise.

3.3 Varieties of Quant UXRs

We divide Quant UX researchers into four general categories: social scientists, survey specialists, metrics specialists, and large-scale data modelers. In practice, most Quant UXRs have skills from two or three of those categories, but it is impractical to expect deep skills across all four categories.

Social scientist Quant UXRs focus on the definition, assessment, and implications of relatively complex cognitive and behavioral *constructs*. For example, the question, "How many users are worried about data security?" requires clarifying the underlying assumptions (Who is a user? What data? What kind of security?). This is followed by work to operationalize and measure the construct of *worry* in the context of data security. Social scientist UXRs often have backgrounds in psychology, political science, and similar fields. The most crucial skills are human subjects research design, statistics, and assessing experimental reliability and validity (see Section 4.3).

Survey specialists field, analyze, and report short-term, one-off surveys to answer important questions and develop long-term platforms for user assessment. At Google, an example of such work is a Happiness Tracking Surveys (HaTS). HaTS allows teams to implement short surveys and track user sentiment inside a wide range of Google products [95]. Chapter 8, "Customer Satisfaction Surveys," discusses such surveys in depth. We use the term "specialist" here to distinguish this role from the more formal role of a *survey scientist*, someone with a degree or other extensive training in survey methodology (see Section 3.4.3). Is this confusing? You're not alone! The distinction commonly confuses hiring managers. The key thing to remember is that many Quant UXRs do specialized survey work although they are not necessarily survey science experts.

Metrics specialists conduct research to analyze user behaviors in a product, with a particular focus on measuring user experience outcomes. This may be for tracking progress toward goals over time, or for comparing design alternatives in an A/B test (see Section 5.3.4). For example, a metrics specialist may be involved in creating a specific definition of what should count as a successful search interaction from a user perspective, and then implementing metrics to evaluate whether a new candidate design for the presentation of search results is an improvement over the existing design. This work may involve product logs analysis, database analyses, and dashboards or other forms of data visualization. These Quant UXRs come from the largest array of backgrounds, and bring many different sets of skills to such analyses, especially data visualization, database skills, and research communication. In Chapter 7, "Metrics of User Experience," we present the HEART framework, which is a helpful starting point for teams to use when defining user experience metrics.

Large-scale data modelers engage in user-centered research using massive data sets, such as the product logs of search engines, social media products, advertising platforms, operating systems, and other products with millions or potentially billions of users. Common questions in this area involve the real-world patterns of product usage, interactions among them, and their relationship to other behaviors of interest. For example, a modeler might ask, "If a user does X in the product, what else are they doing? How does that relate to whether they continue using the product?" Researchers using large-scale data often have backgrounds with an emphasis in both computer science and statistics, such as a master's or PhD in data science or computer science. Compared to a metrics specialist, these researchers tend to do more statistical modeling with usage data, rather than reporting of descriptive statistics. Crucial skills in this area are general-purpose programming, statistics, and database analysis. We discuss those skills in Chapter 5, "Statistics," Chapter 6, "Programming," and Chapter 9, "Log Sequence Visualization."

Of the authors, Chris's work aligns predominantly with social science and survey specialization, while Kerry's work aligns with large-scale data modeling and metrics. Yet each of us has done work in all four areas.

Which area should you emphasize? It depends on your skills and interests! Beyond that, we have one suggestion: in recent years we have observed that the *survey* area is in particularly high demand and often outstrips the available skills of Quant UXRs.

Why is survey research so popular for Quant UXRs? Oftentimes the fastest and most effective way to learn something from users is simply to ask them (in a valid way), and there are many important aspects of user experience that require directly asking and cannot be captured through observation or logs alone. Also, advanced survey methods such as structural equation modeling, psychometrics, and stratified sampling are a good fit for the statistical knowledge of many Quant UXRs. However, many Quant UXRs come from academic disciplines such as computer science, cognitive science, neurobiology, or math, where there is little preparation for doing effective, well-designed survey research. One way to develop expertise is through summer courses offered by university survey methodology programs (see Section 3.7).

3.4 Quant UXR Differences from Other Roles

The role of a Quant UX Researcher overlaps with other roles in its emphasis on research design, empirical research, statistics, data visualization, programming, human subjects research, database skills, user experience, and research communication. What makes Quant UXRs unique? It is the *combination* of skills—Quant UXRs need to have basic competence across all of those areas, whereas analysts in other roles have different combinations of skills.

Although the differences are inexact and often a matter of emphasis or organizational structure, there are some general differences between Quant UXRs and other roles. In this section, we describe the differences as we see them, from the point of view of individual researchers and their week-to-week work. Chapter 11, "UX Organizations," presents a complementary organizational view.

3.4.1 Quant UXR vs. General UXR

In our view, Quant UXRs and general UXRs are closely related. Both roles are UX specialties and typically work on the same or closely parallel teams where they often partner on projects (as discussed in Chapter 11). This is reflected by the authors' experience. Kerry developed the Quant UXR role from the general UXR role. Chris started as a general UXR, moved into Quant UXR, and continues to use qualitative methods alongside quantitative work in unified research.

Despite the similarities, there are two specific differentiators between general and quantitative UXR roles. First, Quant UXRs generally have greater expertise in statistics, programming, and working with data in many different formats. General UXRs may have stronger knowledge of the field of human-computer interaction (HCI) and more experience with small sample research such as usability lab studies and ethnography.

Second, the two roles emphasize slightly different sets of stakeholders. Although general UXRs and Quant UXRs both work closely with product managers, they diverge in their patterns of engagement with other roles. General UXRs often work closely with UX designers, whereas for Quant UXRs it depends on the project. Quant UXRs often have a broader, more diverse set of stakeholders, which may include more direct engagement with executives.

3.4.2 Quant UXR vs. Mixed Methods UXR

We have seen a recent trend in UX researchers describing themselves as "mixed methods" UXRs, and hiring managers looking for such candidates. There is no single definition of *mixed methods* but it implies that one engages in both qualitative and quantitative research.

In our observation, mixed methods UXRs have some of the following characteristics. They may have deep experience with statistics and wish to use that knowledge alongside traditional qualitative methods. A mixed methods UXR may work as a generalist selecting among methods that range from field research and in-person user studies to statistical modeling. Or, mixed methods UXRs may work in an organization where programming is seen as a crucial aspect of the Quant UXR role, but they do not have the relevant experience in coding.

We propose that the difference is this: Quant UXRs should be experts across a relatively broad range of empirical methods and statistics, whereas mixed methods UXRs use such methods opportunistically, according to their individual knowledge and the demands of the project. As we discuss in Chapter 6, basic programming skills are highly desirable for Quant UXRs, but they are not generally a requirement for mixed methods UXRs.

3.4.3 Quant UXR vs. Survey Scientist

The difference here should be relatively obvious: a survey scientist is deeply expert in survey design, methodology, sampling, and statistical analyses for survey data, whereas a Quant UXR may not have such deep expertise in survey methods (although survey work may be part of the job, as noted in Section 3.3). A survey scientist typically has a master's or PhD from a program in survey methodology or closely related fields. Like general UXRs, survey scientists may not have deep experience in programming, statistics, or working with diverse data formats.

There are two important organizational differences between Quant UX and survey scientist roles. First, survey scientists might be located outside of a UX team or product engineering. They are often staffed under marketing, operations, or other non-engineering organizations. Second, a Quant UXR typically engages in a variety of diverse primary research projects to inform ever-changing questions from stakeholders. A survey scientist is more likely to own longer-term projects, often managing external suppliers, and may have less emphasis on primary research.

3.4.4 Quant UXR vs. Marketing Researcher

Marketing research is another area that overlaps with Quant UX research. For example, in Table 3-1, there are questions about willingness to pay, the value of features, and customer segmentation—each of which is traditionally associated with marketing research. While working in a formal Quant UXR role, Chris has chaired marketing conferences and coauthored marketing research books [25, 127].

It is important to distinguish *marketing research* and *market research*, terms that are often incorrectly interchanged. *Market* research—commonly observational and descriptive—concerns *markets*, venues for the transaction of goods, services, and money. Market research may examine which markets—perhaps countries, customer groups, or distribution channels—are most appealing for a product. It asks who the competitors are in a particular market, and what are the important trends for consumer demand.

Marketing research is a larger category that includes many other activities such as experiments to determine consumer response to advertising, product promotion, and other interventions; investigation of brand and product awareness, perceptions, and desirability among consumers; product optimization such as feature definition; and pricing. This involves investigation of individual behavior, often using social science or econometric methods, alongside observation of markets as such. This overlaps highly with the social science aspects of Quant UXR (see Section 3.3).

Perhaps the most salient distinction between marketing research and Quant UXR is the *organizational* difference in technology companies. In tech, marketing is often organizationally distant from engineering and may not engage directly with product definition. Quant UX research, on the other hand, is usually embedded in (or closely aligned with) engineering, where it provides direct input to product definition and engineering leaders. This is not usually the case in other industries. In many consumer goods companies, manufacturing firms, service providers, and other organizations, marketing research is closely involved in product definition. In those organizations, marketing research performs many of the activities that are identified with Quant UXR in Table 3-1 [27, 28, 85].

Overall, we believe that Quant UXRs often have the required skills to adopt and benefit from quantitative marketing research methods [28], yet the two roles remain organizationally distinct in tech companies. In the future, we hope that there will be more organizational awareness and interaction between these roles.

3.4.5 Quant UXR vs. Data Scientist

Now we get to the largest overlap of all: Quant UX and data science. Data science is not a well-defined field, and in this book we will not propose yet another definition, except to note that data science involves a combination of domain knowledge, research design, data acquisition, statistical computation, and data visualization in order, as Wickham and Grolemund write, "to turn raw data into understanding, insight, and knowledge." [149] In fact, that is true of all kinds of scientific endeavors.

What distinguishes Quant UX research from data science? One answer is that it is a subset relationship: Quant UX research is a subset of the much larger field of data science. One might say that Quant UX is the intersection that occurs when data science skills are applied to UX research problems.

We believe that, although this is technically true, it is misleading because it underplays the value of relevant domain knowledge. Quant UX research differs from data science because it has a particular emphasis on *research design* and *primary research*, with *human subjects*, in *UX organizations*. None of those is generally true of data science, yet they dramatically affect the character of Quant UX work.

This set of differences has several implications. First of all, one should not assume that a randomly selected data scientist will have expertise in Quant UX research. Expertise in conducting primary research with humans is a specialized skill. On the other hand, although data science skills are an important part of the toolkit for Quant UXRs, they are necessary but not sufficient.

Closely related to the difference in skills is a difference in how each role influences products. Data scientists often engage in deeply technical projects or implement *production* systems such as reporting pipelines. *Production* describes a system or process that is directly exposed to external or internal users. Its operation is a core part of a product or business, and may perform crucial data processing functions. For example, code that applies a machine learning model and returns results to users is a statistical model *in production*.

On the other hand, Quant UXRs most often engage in the design and implementation of shorter term, focused research, and their code does not usually go into production. A model of user behavior to inform strategy is not a production system. The boundary may be blurred for internal products such a dashboard.

In terms of background, a Quant UXR role is more likely to draw on formal training in research design, as one might obtain in graduate study. Data science skill may be acquired in more diverse ways.

3.4.6 Quant UXR vs. Business or Product Analyst

Business and product analyst positions often overlap with data science (see Section 3.4.5). However, in practice we have observed the following differences from Quant UXR and data science.

First, projects taken on by an analyst may involve fewer aspects of problem definition, research design, and statistical analysis. An analyst may work with relatively clean data whose purpose and need have been previously established. Such data may include user activity, engagement time, revenue, sales within a product or region, and so forth.

The work product of an analyst is more likely to be regular and periodic, such as a monthly report of trends, breakdowns of usage or sales by country, or a metrics dashboard. Analyst roles often emphasize clear, timely, and accurate reporting more than primary research design or statistical analysis.

There are differences in the toolset and skills of analysts. Whereas a data scientist or Quant UXR may use R, Python, or Julia, and write a substantial amount of analytic code, a product or business analyst is more likely to have expertise in platforms such as Microsoft Excel, Tableau, or Google Analytics. Custom code for analysts is often in SQL or a platform-specific language such as Visual Basic for Applications (VBA).

Analysts often report to different parts of the management hierarchy. Whereas Quant UXRs report to UX and engineering, analysts are more often based in dedicated analytics teams within marketing, sales, support, operations, or finance.

Having outlined the differences, we would emphasize again that there is much overlap, and someone with Quant UXR skills might enjoy an analyst position. To pose a single question, ask yourself whether you prefer to work with more nebulous, open-ended, primary research problems centered on users—with a substantial chance of any project failing—or you prefer more structured, clearer projects where success consists of improving tools, processes, and reports used by executive stakeholders. The former aligns better with Quant UXR, whereas the latter aligns with analyst positions.

3.4.7 Quant UXR vs. Research Scientist

Some tech companies designate roles for *research scientists*. Microsoft Research and, more historically, organizations such as Bell Labs and Xerox PARC (Palo Alto Research Center) are examples. In such groups, a researcher may operate much more like an academic researcher, pursuing projects that are not necessarily related to any specific product team or ultimate business outcome.

Although such positions are appealing, there are relatively few of them in industry. A trend in recent decades has been toward embedding researchers in product teams rather than autonomous research groups. Also, in research groups there are few positions related to social science, human-computer interaction, statistics, and similar areas; instead, they emphasize computer science. If you are the kind of candidate who might fit such a role, you probably already know it from association with a specialized, industry-connected academic program.

3.4.8 Quant UXR vs. Academic Research

Many academic social scientists excel at the skills needed for Quant UX research, including deep knowledge of human behavioral research, and skills in statistics and programming. There are obvious differences, such as the setting, compensation, locations, and potential tenure, but we would like to reflect on a few other issues that are often of concern for academics.

One question is the origin of research projects. In our experience, it is usually not the case that Quant UXRs are "given a project" or told exactly what methods or analyses to use. Instead, Quant UXRs apply their own skills and interests to select among possible projects, design appropriate research, and lead stakeholders to action. In some ways, this resembles what the principal investigator (PI) on a funded research grant might do: take a general problem area identified by the funding agency, propose research, and see it through to completion. Quant UXRs are PIs who are "funded" by the organization where they work.

The audience for Quant UXRs differs from that of academics. Quant UXRs present research primarily to non-specialists who may be designers, engineers, marketers, or business executives (see Chapter 13). Quant UXR presentations should be less technical than academic presentations and focus on business and product recommendations. These research reports receive less scrutiny and technical examination than academic work, which has both positive and negative aspects.

Academics often wonder about industry policies for publication and external engagement. Policies differ across companies, yet our common experience is that publications and teaching are passively supported as side projects. As long as there is no sharing of sensitive, proprietary information or algorithms, and no release of private data, many firms will approve publication of research, serving on conference committees, and the like. They may even cover associated costs, such as attending conferences or paying a journal for open access rights. On the other hand, publishing is unlikely to bring short-term career rewards such as better performance reviews, higher compensation, or promotion. Although firms regard publication as good in an abstract way, it is much less important than week-to-week work.

Overall, if someone wants to teach or publish as a UX researcher, there is usually a way to do that. Although it will likely have little impact on a career in the short term, in the longer term it can be highly beneficial. Through teaching one meets many people and gains a reputation as an expert. By publishing or presenting at conferences, one is required to improve skills and learn from others. Those experiences pay off in multiple ways over the course of years or decades, and ultimately will help your performance reviews and career. Thus, we strongly encourage such activity; the trade-offs are among your own time, energy, and enthusiasm along the way.

3.5 Will You Like a Quant UXR Role?

As the previous sections demonstrate, there is a great deal of overlap between Quant UX research and other roles such as data science. The following questions might help you to sort out some of the distinctions among general UXR, Quant UXR, data scientist, and analyst. To be sure, the questions do not form a validated assessment instrument! We suggest them as a simple way to keep score for purposes of self-reflection.

Grab a piece of paper and jot down entries for (1) General UXR, (2) Quant UXR, (3) Data Scientist, and (4) Business or Product Analyst. Then answer the following questions and score your responses.

1. *Are you strong at statistics*, at the approximate level of an undergraduate statistics major or a graduate social scientist?

   - *Yes*: Score 1 on the lines for both Quant UXR and Data Scientist.

   - *No*: Score 1 on the lines for both General UXR and Business or Product Analyst.

2. How strong are you at programming?

- *Level 0*: Programming is new, uncomfortable, or not enjoyable. Score 1 for General UXR.

- *Level 1*: You enjoy writing scripts such as basic R or SQL code. Score 1 for Business or Product Analyst, and score 0.5 for Quant UXR.

- *Level 2*: You often program in a general-purpose language such as R, Python, C/C++, Julia, etc. You routinely write hundreds of lines of code, and think in terms of reusable functions, replication, and formal ("unit") testing. Score 1 for both Quant UXR and Data Scientist.

- *Level 3*: You have a computer science degree or the equivalent, with deep experience in algorithms and formal systems, and may have worked as a software engineer. Score 1 for Data Scientist and also score 0.5 for Quant UXR.

3. Do you have *experience designing and leading human or behavioral research*, such as prior work as a human subjects researcher, or completion of a social science PhD?

- *Yes*: Score 1 for both General UXR and Quant UXR.

- *No*: Score 1 for both Data Scientist and Business or Product Analyst.

4. Do you have any of the following *graduate degrees*?

- *MBA or other business graduate degree*: Score 1 for Business or Product Analyst.

- *Statistics*: Score 1 for Data Scientist and 0.5 for Quant UXR.

- *Human-computer interaction or human factors*: Score 1 for General UXR and 0.5 for Quant UXR.

- *Computer science* (other than HCI and human factors): Score 1 for Data Scientist.

- *Psychology, sociology, or political science*: Score 1 for both General UXR and Quant UXR.

- *Neuroscience*: Score 1 for Quant UXR and 0.5 for Data Scientist.

- *Anthropology, ethnography, or information science, other social sciences*: Score 1 for General UXR.

- *Natural science or mathematics*: Score 1 for Data Scientist.

- *None of the above*: Score 1 for Business or Product Analyst.

5. Do you have *UX work experience*, such as any role in UX research, HCI, interaction design, or UI design (such as web design or application front-end design or programming)?

   - *Yes*: Score 1 for both General UXR and Quant UXR.

   - *No*: Score 1 for both Data Scientist and Business or Product Analyst.

6. Which of the following *types of projects* would you most prefer? (Choose one.)

   - *Working with a designer to create the UI of an app*: Score 1 for General UXR.

   - *Developing a data pipeline to process user-generated feedback and use it for product ratings*: Score 1 for Data Scientist.

   - *Examining trends in customer satisfaction ratings or user engagement metrics, and determining what they mean for the next version of a product*: Score 1 for Quant UXR.

   - *Reporting trends in product sales to executives and regional marketing teams*: Score 1 for Business or Product Analyst.

7. *Still unsure?* Check the skills and project examples later in this book. Then apply to your top two possible roles, and use the interview process as a way to learn more. (See Chapter 12, "Interviews and Job Postings.")

In the appendices, we provide an example of a job description (Appendix A) and hiring rubrics (Appendix B) for a hypothetical Quant UXR job position.

3.6 Key Points

In this chapter, we gave a general definition of Quant UX research, described a Quant UXR's typical activities, and contrasted Quant UX research positions with other roles such as data science. The following are key points covered in this chapter:

- Quant UX research is the application of empirical research methods to inform user-centered product design at scale (Section 3.1).

- Research questions and methods vary across the product lifecycle as stakeholder and product needs change (Section 3.2.2). It is typical to work simultaneously on shorter- and longer-term projects that span multiple parts of the lifecycle (Section 3.2.1).

- There is no single set of skills for Quant UXR. A researcher might specialize in one or two areas such as social or psychological constructs, behavioral metrics, survey research, or large-scale data modeling (Section 3.3).

- Quant UX overlaps significantly with several other roles, including general UX research, data science, and product and business analytics (Section 3.4).

- We believe that Quant UX research has the highest appeal to those who enjoy tackling nebulous, human-centered research questions with a combination of experimental research design, programming, and statistics (Section 3.5). This may include candidates from academia, data science, HCI, statistics, social science, and other domains.

3.7 Learning More

For social scientists, and especially psychologists, Chris has elsewhere described how skills in human research are advantageous for the activities and needs of general UX research [17]. For statisticians, Tim Hesterberg has outlined common projects, job opportunities, and ways to learn more about roles for statisticians at Google [59].

If you conclude that your interests align best with general UX research, an excellent guide to descriptive and basic inferential statistics for UX is Sauro and Lewis, *Quantifying the User Experience* [124]. Note that Quant UX roles expect a higher level of skill in statistics and experimental design than their book describes.

A Quant UXR may be interested to develop survey research skills. Several universities in the United States and Europe offer summer programs that last from one to six weeks and offer immersive training in survey research methodology. In recent years, there have been summer schools in survey methodology and related areas at the University of Michigan (United States); the GESIS Leibniz Institute for the Social Sciences (Germany); Pompeu Fabra University, Barcelona (Spain); the University of Essex (United Kingdom); and Utrecht University (Netherlands).

A recurring theme in this chapter is the overlap in skills and projects between Quant UXR and data science. Robinson and Nolis provide guidance from the data science perspective in their book *Build a Career in Data Science* [114].

In Table 3-1, we noted several common methods used by Quant UXRs. Each of those methods has a deep literature and it is easy to get lost as a newcomer. The following references are particularly useful as starting points for applied projects:

- *Logs analysis*: There are many kinds of logs analysis. In UX we most often examine user *session logs*, sequences of user behaviors in an application or website. An introduction to sequence analysis is provided in Chapter 9 of this book.

- *MaxDiff choice modeling*: MaxDiff is a forced-choice survey method that is helpful to assess the relative importance of features, use cases, and product messages. See Chapter 10 in this book for an introduction to MaxDiff.

- *Conjoint analysis*: An introduction to conjoint analysis in R is given in the R companion text [25]. Bryan Orme has written a less technical introduction with extensive guidance for practitioners and research managers [106].

- *Brand perception*: A popular approach to assess competitive brands' perceptions is *composite perceptual mapping* (CPM). CPM is demonstrated in Section 8.2 of the R companion book [25], and Section 9.2 of the Python companion [127]. Another approach is *sentiment analysis* of open-ended text, such as product reviews and user responses to surveys. See [132] for an excellent practical introduction.

- *Clustering (aka segmentation)*: *Cluster Analysis* by Everitt, et al. [43] is a classic and approachable introduction to cluster analysis. Examples of clustering using R and Python are given in [25] and [127], respectively. The most important point we urge researchers and stakeholders to understand about clustering is that it is predominantly an *exploratory* method that requires an analyst to apply a large amount of judgment. Clustering is not a matter of simply applying a statistical method to a data set to get "the best answer."

- *Classification*: Classification is covered in many statistical textbooks, among which we especially recommend Kuhn and Johnson's *Applied Predictive Modeling* [76]. Its associated R package `caret` provides structured access to hundreds of classification models (238 models as of this writing) [75]. The R [25] and Python [127] companions also discuss classification.

- *A/B and multivariate experiments*: Similar to classification, experiments are covered in a general way in a vast number of statistical texts, because they are variations of classic inferential testing. However, there are many potential pitfalls and issues that are not discussed in general texts. For analysts using R in business settings, McCullough's *Business Experiments with R* [92] is an excellent and readable guide that presents applied examples and discusses best practices.

- *In-product and online surveys*: A common mistake with surveys is asking what you want to know—instead of what respondents are able to tell you [20]. Too many surveys are not planned well, written well, or pre-tested. Jarrett's *Surveys That Work* [63] is a practical guide that emphasizes careful consideration of the decisions you might take according to survey results, and then designing appropriate research plans and items to inform those decisions. More detailed guidance and a comprehensive summary of empirically-based recommendations is Callegaro et al., *Web Survey Methodology* [12]. If you do many surveys, we recommend that you learn about scale development and psychometrics generally; DeVellis's *Scale Development* [39] is an outstanding, approachable introduction.

- *Experience sampling*: These methods range from almost exclusively qualitative approaches, such as diary studies, to longitudinal structural models. A good introduction is Silvia and Cotter, *Researching Daily Life* [133]. Bolger and Laurenceau [7] discuss technical methods for analysis of longitudinal within-person events and responses, with code for many examples available in R as well as SAS, SPSS, Mplus, and HLM.

- *Causal modeling (causal inference)*: Every analyst is confronted with stakeholders who interpret relationships causally, confusing causation with correlation. ("User satisfaction dropped? What else changed? That must be the cause!") Statisticians have developed models that are able to assess potential causation. A fun introduction with just enough math, and code in R and Stata, is Cunningham's *Causal Inference: The Mixtape* [38].

This list of skills should not scare you! No Quant UXR can be an expert in all of them. Rather, one becomes progressively more expert in a few of them in the course of a career, while learning the basics of many others (see Section 4.1.1).

One of the great things about Quant UXR is that there is always much more to learn.

PART II

Core Skills

Introduction to Part II

In Part II we examine the three essential and somewhat unique sets of skills for Quant UX: research design (Chapter 4), statistics (Chapter 5), and programming (Chapter 6). Our goal in this part is to *define* the skills needed. We do not attempt to teach all of the skills, because each topic is vast. Instead we describe why each skill is important, outline its boundaries, discuss expert-level skills that are *not* required, and recommend areas for self-assessment and learning.

These chapters will be of greatest value to three audiences: readers who wonder how their own skills compare to those needed for Quant UX, readers who are considering applying to Quant UX roles and wonder what they need to know, and UX managers and stakeholders who wish to create Quant UX roles or help team members to develop their skills.

Some readers will be reassured that their skills align well with Quant UX. Others will uncover areas for additional learning. And still others may conclude that Quant UX is not a good fit for their skills or interests. Each of those is a valuable outcome.

CHAPTER 4

UX Research

Quant UX research requires skills in several areas, including research design, statistical analysis, and (usually) programming. In this chapter we take a look at the Quant UX approach to research design.

Our central claim is that the most important aspect of research skill for Quant UXRs is *thinking like a UX researcher*. We focus on this for a few reasons. First, we believe this approach is more important than specific technical knowledge. If you get the approach right, you can then figure out the details. Second, the range of possible technical knowledge is so vast that no one can possibly cover it all. Third, technical documentation for analyses are readily available, but there is much less written about the "how and why" of Quant UX practice.

This chapter discusses several underlying user-centric concepts that apply to both qualitative and quantitative UX research. We do not go deeply into specific methods, either quantitative or qualitative, but instead present an organizing set of principles. Where applicable, we use examples that might arise in Quant UX research.

By the end of this chapter, you will understand much about how UXRs and Quant UXRs think about research problems. This will help you formulate appropriate research plans for Quant UXR projects (and interviews) and further understand differences from related fields such as data science (Section 3.4.5).

As you read this chapter and the next two chapters on recommended skills in statistics and programming, it may help you to refer to the condensed set of hypothetical hiring criteria in Appendix B.

4.1 Foundational and In-Depth Skills for Quant UXR

There are three areas of specific technical expertise that are typically emphasized for Quant UXRs: statistics, programming, and UX research (or, more broadly, human-computer interaction research). We discuss statistics in Chapter 5, "Statistics," and programming in Chapter 6, "Programming." Here we focus on UX research.

© Chris Chapman and Kerry Rodden 2023
C. Chapman and K. Rodden, *Quantitative User Experience Research*,
https://doi.org/10.1007/978-1-4842-9268-6_4

The relationship among the high-level skills for Quant UXRs is shown in Figure 4-1, based on Drew Conway's data science Venn diagram [34]. Quant UXR sits at the intersection of the three skill sets of programming, UX research, and statistics.

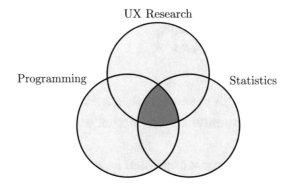

Figure 4-1. *A Venn diagram with the three principal skill sets that are required for Quant UX research: programming, UX research, and statistics. They intersect to form Quant UX research at the center*

Although Figure 4-1 is simple in structure, there is an important implication: depth in statistics and programming is not enough. Domain expertise in UX research is just as important as those technical areas. In fact, UX research skills are also technical, and you should not underestimate the value of expertise gained through deep study and experience in the domain.

An immediate practical aspect of Figure 4-1 is this: if you apply to Quant UX jobs, you should expect to be interviewed in each of these areas. A typical interview process will have four to six areas that are assessed for competence above a *minimum bar*. For Quant UXR, this usually includes assessment of statistics and UX research design, often includes programming, and typically adds areas that are common to all professional jobs (such as communication skills) or are required for a particular position (such as specific language or database experience). We say more about hiring processes in Chapter 12, "Interviews and Job Postings."

4.1.1 "T-Shape" Skills

In addition to the breadth of skills that are required at a basic level (see Figure 4-1), every Quant UXR should have demonstrable *expertise* in at least one area. This is sometimes known as a *T-shaped distribution*. The breadth of skills is represented as the horizontal top bar of a *T*, while depth in one area is represented by the descending bar.

Figure 4-2 illustrates a T-shaped set of skills where a person has basic proficiency in programming and UX research along with in-depth expertise in statistics.

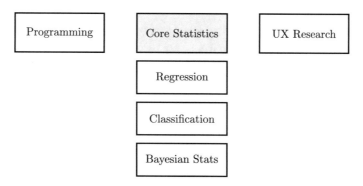

Figure 4-2. *An example of the T-shaped relationship among skills for Quant UXRs. Every Quant UXR should have a set of broad, basic skills that surpass a minimum requirement (visualized as the top of the T) along with specific depth well above the minimal requirement in at least one area (shown as the tail of the T). This figure illustrates breadth in programming, statistics, and UX research, along with in-depth expertise in statistics*

The T shape may be used to describe an individual's skills, or a particular Quant UXR position that requires depth in some area that depends on the team or organization. Important questions for you to consider are

- Do I have basic competence in all of the breadth areas?

- In which areas do I have in-depth expertise?

- How do I communicate that expertise?

- For career development, would I prefer to develop additional breadth, or instead to develop more depth in my specialty areas?

We say more about those questions in Chapter 12 and in Chapter 14, "Career Development for Quant UX Researchers."

4.2 Focus on the User

In its early years, Google developed a list of "Ten things we know to be true" as a reflection of its organizational philosophy. The first item became probably the best-known: *"Focus on the user and all else will follow."* [51] Because we both worked at

Google for many years, we can't help but recall this statement when we think of the purpose of UX research and design. In our view, this is the most important thing that a UX researcher can do: focus on the user.

In this section, we'll expand on what it means to focus on the user, and how that differs from other approaches. We use the term *user-centric* to describe this. In our view, there are five key principles in a user-centric approach to research:

1. Adopt the user's perspective.

2. Assess user-centric variables and outcomes.

3. Answer "why?" with a cognitive approach.

4. Focus on unmet needs.

5. Relate to UX actions and stakeholders.

In the following subsections, we examine each of these principles.

4.2.1 Adopt the User's Perspective

The first principle is to adopt the user's perspective. UX research questions should focus first on how a user would consider the problem or product.

Here's an example, inspired by a project where Chris led the UX research [26]. Imagine your team is investigating a digital writing device: a "digital pen" that records what a user writes and delivers the handwriting as an electronic document. Reasonable business questions include the following:

- Should your team make this product?

- What are the key engineering challenges?

- Can your team overcome the engineering challenges?

- How much market share will your product gain?

- Will the product be profitable?

But none of those is a *user-centric* question. When you consider the proposed digital pen from a user's point of view, you ask very different questions:

- How does a digital pen help a user?

- Why is it better than traditional pen and paper?

- How will people use it in real situations?

- What proportion of users' writing tasks can it accomplish or improve?

- What happens if a user doesn't have it, it's not charged, it breaks, it's lost, and so on?

- Given the advantages and disadvantages, is it worth the cost?

Each of those user-centric questions also helps to answer a business question, because if users don't see value in a product, then the business selling it will struggle. The goal of UX research is to bring in the user's perspective to help ensure that engineering and business directions are best aligned with what users really need. If your research plans have more to do with the *business* than with the *user*, then you should consider how user-centric research can bring additional perspectives.

By the way, the digital pen research by Chris and colleagues (in the 2000s) found that traditional, non-digital pen and paper solutions are cheap, adaptable, ubiquitous, and universally familiar, and have been optimized over thousands of years. From a user's perspective, the behavioral value of traditional writing is a huge barrier to the widespread adoption of digital writing [16, 26].

4.2.2 Assess User-Centric Variables and Outcomes

A closely related principle is this: assess user-centric variables and outcomes. Quant UX researchers should design studies and work with data sets that relate *human* (user) data to human outcomes. This often takes the form of a statistical analysis where the inputs (sometimes called predictors or independent variables) come from user actions and are related to user-centric outcomes.

For example, assume we have detailed behavioral data about users' interactions with a website, such as their visit frequency and duration, as well as survey data collected directly from a subset of users about their satisfaction with the site. We could use the behavioral data as the statistical predictors (inputs) and the satisfaction data as the dependent or outcome variable, and develop a model to help us understand which user behaviors are most related to satisfaction. That understanding would help us to develop behavioral metrics to use in A/B tests (see Section 5.3.4) when evaluating new designs.

Common user-centric data include the following: user behaviors, desired tasks, task success, product adoption, feature usage, preferences, environments (e.g., home or office setup), survey responses, purchases, returns, product reviews, survey comments, support requests, bug reports, and subjective understanding. Most commonly, any of these would be an outcome variable to be related to a treatment such as product release, or a factor such as user status or demographics. However, depending on the research question, any of these might also be regarded as a predictor variable or covariate of some other variable.

What are variables that are *not* related to users—or, to put it more precisely, are not closely or directly related to individual users? On the input/predictor side, some examples include geographic region; engineering details that are hidden from a user; organizational variables such as assignment to an internal segment, engineering team, sales team, or goal; impersonal data such as economic trends, political events, weather; and so forth. In terms of outcome or dependent variables, examples of non-user-centric data include revenue, profit, market share, expert reviews, news articles, staffing (personnel) levels, system errors, data center demand, energy usage, manufacturing details, and many others.

Of course, all data sets are *indirectly* related to users. What's the difference from user-centric data? For a particular set of data, ask yourself whether a user could reasonably *say something* about it from their own perspective and, if so, whether you would really care what they said. If the answer is "yes," then it's user-centric data. If not, then it quite likely lies outside the domain of UX research. There are many areas of overlap, but the *center* of the focus is the crucial aspect.

4.2.3 Answer "Why?" with a Cognitive Approach

Understanding *what* users do is incomplete unless we also understand *why* they are doing it. To understand users, we need to understand their thinking as well as their behavior. As UX researchers, we do this by taking a *cognitive approach*.

Like any research project, the details of how to include cognitive aspects will vary according to the question at hand. However, there are several common considerations for Quant UX projects:

- Have our data sources or questions been reviewed with actual users? Too often, we collect data that doesn't mean what we think it means. Consider the common survey question, "How likely would you be to recommend ___?" Strictly speaking, every answer to that could be close to "never." It is a fact that few of us spend much time recommending anything to anyone. So what does the survey question really mean? Answer: it doesn't mean anything *in itself.* Rather, its meaning comes from how users interpret it. To understand that, you need to ask them.

- When we observe users doing something, do we know *why* they are doing it? It is common to assume reasons that are untrue. Suppose a large number of users report a preference for the Windows operating system, perhaps on a MaxDiff survey (see Chapter 10, "MaxDiff: Prioritizing Features and User Needs"). We might wonder why Windows is so important to them...but that is already the wrong question to begin with. First we would want to understand what they think "Windows operating system" means. Does it mean to them what *we* think it means? Users might believe that every computer has Windows, or that Windows is the same as Microsoft Office. We won't know unless we ask them *why* they made some choice.

- Have we assessed—even at a very basic level—the emotional aspects of users' experiences alongside the behavioral? As a human-centered discipline, we aspire to design product experiences that contribute to *happiness* as well as objective task success. Put differently, if users succeed at using a product but don't *enjoy* it, the product will have a very uncertain future and may be a failure. In Chapter 7, "Metrics of User Experience," we describe the HEART model [118] for multidimensional assessment of behavior and attitudes.

4.2.4 Focus on Unmet Needs

Engineers, designers, product managers, and technology executives love innovation. One of the best things about working in UX is to be involved with fascinating new product ideas.

Unfortunately, the focus on innovation leads to a common problem in UX research: asking about how users respond to the new product concept. This takes the form of asking whether they want the product, whether they will use it, what they need it to do, how much they will pay, and so forth. Those are good business questions, but we hope by now you will see the problem: they are not especially *user-centric* questions.

The question for a user is likely to be a variation of, "What will it do for me?" That includes the implicit question, "Why is it better than the alternatives I have now?" We saw an example of this in the discussion of digital writing in Section 4.2.1.

A good place to start for UX research is to learn about users' *unmet needs*. What problems do users have with current solutions? (These are sometimes called "pain points," although we adopt a broader approach.) How serious are those problems? What are users not able to do because of the problems? How valuable is a solution to them? What are they doing now, which might change? And then, does our proposed solution actually *address* those unmet needs? It is often helpful to begin such exploration with qualitative research. Direct observation of users is very effective at uncovering problems and revealing things you otherwise wouldn't think to ask.

If this seems obvious to you...good! But we can attest that it is often forgotten in the world of product development. A telltale sign is when a team talks about what *it* needs from users—such as wanting users to engage more or buy more—and especially when that includes focus on a competitor. The history of technology is full of cases where a company observed success of a rival and responded, "We need to get into that market, too. We can't let them have all the success!"

If the reason we want to develop something is because it will be better at meeting users' needs, there is no problem. But if we are unclear on the users' unmet needs, and are focused on figuring out how to make users want our product, then a far more likely outcome is a treadmill of poor ideas that users don't want, with repeated research studies hoping to uncover a magical insight.

A classic paper by John Gourville, "Why Consumers Don't Buy" [52], describes how product teams greatly overestimate the value of their offerings. At the same time, consumers underestimate the value and wish to avoid possible disappointment and the "switching costs" of changing behavior. UX research can help with both sides of that balance, acting to align teams' estimates of the value with real user needs while also understanding the actual barriers that users may expect (even if the barriers are unreal or exaggerated).

Our recommendation: focus consistently on what users need, what their current solutions don't do, and what the behavioral and cognitive implications are for any new solution. Design for real unmet needs, not hypothetical benefits.

4.2.5 Relate to UX Actions and Stakeholders

The final principle is that Quant UX is part of UX. Our research problems typically come from UX stakeholders such as designers, PMs, engineers, executives, and UX colleagues in other roles. UXRs recommend actions that UX stakeholders are able to implement.

Some common recommendations in UX involve changes to the user interface (to be considered by designers), new content to include or extraneous content to remove, observations of unmet needs (to be prioritized by PM, design, or others), and descriptions of target user groups. We won't belabor the point, because all of this book is about how Quant UX relates to UX more generally. Key considerations include the research lifecycle (see Section 2.2.2), how Quant UX relates to other roles (see Section 3.4), UX organization structures (see Chapter 11, "UX Organizations"), and working with stakeholders (see Chapter 13, "Research Processes, Reporting, and Stakeholders").

4.3 Research Validity

In your exposure to research design, you have likely encountered the general concepts of research *validity* and *reliability*. Briefly, validity concerns whether research is actually about what it claims to be about, and whether it makes sense on the basis of domain knowledge, prior research, and general logic. Reliability concerns (in part) whether a particular user assessment—such as an observation or a survey score—can be trusted, and whether another assessment would obtain the same results again.

You might consider these to be basic tenets of research. Why do we emphasize them? Because they are often violated in real UX research. Unfortunately, the incentives in industry research too often lead to inadequate research planning, short timelines, excessive optimism, and rewards for delivering what stakeholders want to hear. Simultaneously there is little appreciation of research quality, especially when it delivers an unwanted answer. This situation can trap even the most earnest and well-intentioned researchers unless they take care to remain diligent about research quality.

Two particularly common traps are doing analyses to "see what the data say," and using methods that promise impressive results. The first trap—seeing what the data say—runs into trouble because any set of data will appear to say *something* when analyzed. There are incentives to deliver results and to demonstrate that research delivered something useful. This leads to immense pressure—unconsciously if not consciously—for a researcher to deliver something "of value" to stakeholders. However, unless the data are valid for the problem at hand, and unless the results are reliable, then the most likely value of the research is zero (or negative). Yet few if any stakeholders would recognize that, especially if a project tells them what they want to hear.

The second trap—using methods that promise impressive results—arises through a combination of intense desire for results, enthusiasm to try new methods, and a research ecosystem that is flooded with aggrandized methods.

A classic example in this domain is the Net Promoter Score (NPS), a survey item that asks about likelihood to recommend a product or service. NPS was promoted in *Harvard Business Review* in 2003 in an article titled, "The One Number You Need to Grow." It claimed, "The best predictor of top-line growth can usually be captured in a single survey question." [111] This is an excessive claim because there is no reason to expect that business success could be reduced to a single metric or be captured in a single survey item. It is far more likely that any results that show such a connection are due to random chance, cherry-picked data, fraud, or self-serving exaggeration. Yet it is very tempting to wonder whether it could be true—especially when it is backed by the authority of a prominent publication.

Why not try it and see? The problem is that such effort distracts from more realistic, complex, grounded, and multidimensional research. It misdirects research into a narrow avenue that is insufficient to understand user behavior (see Chapter 7) or make informed decisions. Scholarship soon questioned the validity and excessive claims of NPS [67, 136] but it was too late; the promise was so appealing that criticisms were often ignored. In addition to the discussion here, we say more about NPS and satisfaction scores in Section 8.2.3.2. The point is not that asking about NPS is *bad*, but that you should not entertain inappropriately high expectations for any isolated, simple method.

Such fads come and go regularly in industry, thanks to the pressures and incentives that we previously described. Chris and his colleagues have written about other examples that overpromise and have weak scientific bases, including personas (fictional descriptions of users) and the Kano method (a two-item survey that promises to indicate the strategic priority of product features) [24]. According to the old saying, if it seems too good to be true, it probably is (i.e., it is not true).

As a Quant UXR, what should you do to avoid such problems? First of all, ground your research in logic and prior research. Could you explain the rationale to a skeptic using something more specific than a mere appeal to "trying" it? Second, if you try something new, approach it critically. Ask how you would know whether it is *really* true and not just delivering an answer that you (or a stakeholder) want to hear. Third, build redundancy into your research. Always try to assess a problem with multiple methods, assessments, analyses, or indicators and see whether they agree. Fourth, if you are forced into using a dubious, limited, or discredited method, make the best of the situation by adding some *other* element to the research that is more reliable and valid, or at least that you will learn from. Finally, partner with UX colleagues to add perspectives and increase influence with stakeholders.

4.4 Assessing Users and Assessing Products

We should distinguish two aspects of UX research assessment: assessing *users* vs. assessing *products*. This distinction is often misunderstood by stakeholders and partners, but it is crucial for research planning, statistical analysis, reporting, and general inference about products.

Assessing *users* means that we want to characterize them or their behavior in some way. This relies on traditional statistical concerns of representative sampling, power analysis to be able to find effects or differences of interest, and communication of uncertainty and the distribution of effects across a sample. The key point is that users vary from one another, and we need to understand those individual differences and the distribution of differences.

Assessing a *product* is quite different for the simple reason that a product—for purposes of UX assessment—may generally be conceived to be a single *unchanging* thing. Whereas users behave differently from one another and from themselves at different times, a product should repeatedly respond the same way, given the same conditions. (We are setting aside the specialized area of *operational assessment*, such as measuring the frequency of manufacturing defects or time-to-failure of devices or systems. Those areas are outside the scope or consideration of UX research, where we may assume that a product is fixed at any given point in time. If interested, see Norton [102] for an introduction to metrics for software development.)

What's the difference between assessing users and products? Consider the example of doing UX research for a virtual reality (VR) headset. If we are interested in whether users want it, or how they use it, or what concerns they have, those are assessments of *users* and we will want to report things such as the average interest and the range of interest. We would not particularly care if one user out of ten hates the idea, as long as others are interested.

But now consider the question of physical fit of the VR headset, where we have users try on the device in a usability lab situation. Imagine that it is adjustable and that it fit fine for nine out of ten users...but one user cut themselves while putting it on and was bleeding! In this case, it is not the user who is defective but the product. When a user is bleeding, no confidence interval or report about "average" fit makes sense; we need to fix the design.

Put differently, it takes only one user to uncover a *bug* in a product. The role of UX research in that case is to assess the severity of the bug. Does it occur frequently? What is its significance? Such research may use smaller samples because we are concerned with detection of single events rather than assessment of average outcomes. For more on this, see the discussion of sample size planning in Nielsen [99]. On the other hand, when we want to characterize users and their diversity, larger samples and traditional statistical inference come into play.

4.5 Research Ethics

Ethics is an area of rapid change in the practice of both data science and UX research. In the recent past, many researchers simply assumed that anonymized data could be used for nearly any purpose. Today, there is awareness that even anonymized data may lead to harm to individuals and that there are ethical obligations to society as a whole.

Although we can't review the entirety of research ethics, we would highlight a few areas and issues for your awareness. Depending on your background, some of these will be obvious while others may be surprising.

4.5.1 Research Risks and Benefits

The goal of UX research is to improve products in order to help users, to thereby benefit our business, and ultimately to contribute to the world as a whole. This implies that UX research should have *benefit*, and that any harms it may cause should be greatly outweighed by the benefits. Generally speaking, the risks of UX research are quite low (wasting time, feeling embarrassed, and so forth), but they should be at least informally assessed for any research project.

Quant UX researchers may be most likely to encounter risk of potential harm in connection with A/B tests (see Section 5.3.4), given that such tests involve exposing a large number of users to a change in the product. The change may have real-world negative consequences for those users, depending on the nature of the product. For example, A/B tests in a tax filing product run the risk that some users may inadvertently fail to file their taxes or may receive a lower refund than they otherwise would. It is essential to consider potential harms and mitigations before launching A/B tests that expose users to these kind of risks.

When working with a population at increased risk, such as people with disabilities, or a high-stakes product such as one in a regulated industry, you should consult with a specialist or colleague with experience in the area. When there is substantial risk, detailed disclosures and careful informed consent procedures are important.

4.5.2 Privacy and Legal Requirements

There are two important points here. First, data privacy is becoming increasingly regulated, and the regulations differ by country and transnational region (such as the European Union). You will need to consult on the relevant laws with institutional attorneys wherever you collect data. Second, being *legal* does not make something *ethical*. The law is a minimum requirement, not a definition of ethical conduct. Our ethics should be above the legal minimum.

4.5.3 Minimum Collection

From a research ethics perspective, you should collect the minimum amount of data needed to accomplish the research purpose. Many technology systems capture vast amounts of data for purposes unrelated to research. Setting aside the question of whether a particular system *should* collect such data, we note that the existence of data does not convey an ethical right to use it for *research* purposes. You should consider the ethical situation holistically. Additionally, the existence of vast amounts of data increases the risks of finding results that are invalid or unreliable (see Section 4.3).

4.5.4 Scientific Standards

Quant UX research should conduct itself in accordance with good scientific practice. This means to use appropriate methods, to use them with expertise, and to report results accurately.

We have observed two common patterns where UXRs (and many other analysts) fall short of good scientific practice. The first is in using methods they do not understand well. For example, modern statistics software makes it very easy to fit prediction models, such that any UXR could apply a prediction model to any data set. Suppose a researcher obtains a predictive accuracy of 99% and reports to stakeholders that they have a nearly perfect predictive model. Sounds great, right? Yet have they properly separated training data and its proxies to ensure that the results do not depend on the training data? Have they considered an appropriate baseline? (99% accuracy would be poor if the baseline incidence is 0.001%.) It is not necessary to be an expert in every method that you use, but it is essential to understand common problems and to consult with experts before relying on results.

A second pattern occurs when UXRs attempt to shape their research findings to be useful. Imagine an A/B test of a new design B vs. an existing product design A. Suppose users widely dislike design B and list many reasons for their dissatisfaction. A UXR might report the most important problems with B along with recommendations to improve them.

That seems like a benign research report, right? Perhaps not! Consider the implicit message in such a report: that design B should be taken as the baseline for assessment (because the report is discussing how to improve it), and that the most important decisions for the team are to tweak and improve various details of it. Yet the real message from users might be that design A is far better overall—a message that is buried when the report discusses only granular improvements to design B. In such a case, a well-intentioned research report may end up misrepresenting how users actually responded. That is a disservice to users and, despite seeming immediately palatable and aligned with expectations for product research, it is a disservice to the organization.

4.5.5 Impact on Society

The most difficult question in product research concerns the ultimate impact of our decisions on society. We see every day that consumer products, technologies, and services are changing society. Indeed, the development of human civilization has been directly linked with technology for millennia. Technology products themselves also embody and influence ethical systems [15].

It is tempting for researchers to write off such concerns as "not my job," "above my pay grade," or "impossible to predict." Although we can't solve the problems of global ethics and technology, we can offer three specific recommendations.

First, although some ethical concerns have no exact solutions, that does not imply that insight and improvement are impossible. There are frameworks for thinking about global ethical issues and any researcher should consider learning more about them (see Section 4.8).

Second, remember the first premise of UX research: focus on the user (Section 4.2). As researchers we can engage with users to understand the ethical concerns from their point of view. That doesn't mean asking users to solve the ethical issues, or even to understand them, but rather it means that we should explore the issues with them in a constructive way [18, 19].

Third, if you have ethical concerns about a research or engineering project, you do not have to participate. If a product or research process concerns you, consult with a trusted colleague. Change projects or jobs if needed. If you question whether a product contributes to the net good in the world, don't work on it. There are products in the world that we, the authors, would not choose to work on. That is an individual choice for every researcher, and we don't claim that our choices should be *your* choices. Yet everyone should make choices that feel appropriate to them.

4.5.6 The Newspaper Test

If you are unsure about the ethics of a research situation, ask yourself this question, "How would this look on the front page of a newspaper?" In some situations, this might instead be, "How would my VP or the company CEO respond to this? Would they agree with what I've done?"

If the answer to the question is one of concern or worry, then stop! Get consultation from an experienced colleague—preferably one who works in your organization but on a different product, without a direct reporting structure—and reconsider the options you have. You may need to restructure the research, find other ways to inform a business decision, bring the concerns to the team, or find another project or job.

4.6 Research Planning

Stakeholders and new researchers often expect that the stages of a research project are similar to the following:

1. Receive a research question.

2. Run a study.

3. Present the result.

This model aligns with rudimentary descriptions of the scientific method but is highly incomplete as a description of the actual practice of UX research.

As we emphasize throughout this book, the "research question" usually needs substantial clarification to identify the actual business decision at hand. If there is no decision to be made, there may be no need for research. Similarly, what it means to "run a study" is a complex set of tasks that involves planning, feasibility assessment, and feedback in addition to acquisition and analysis of data.

For Quant UX projects, a more realistic set of steps is the following:

1. Clarify the research question.

2. Identify data sources.

3. Explore possible paths to an answer.

4. Scope the time and effort needed.

5. Review those with the stakeholders.

6. Acquire initial data and pre-test the analyses.

7. Acquire and clean the full data set.

8. Build a statistical model or analysis.

9. Repeat steps 6–8 as needed (for example, if data are bad or a method fails).

10. Frame initial results.

11. Preview the results with a stakeholder.

12. Iterate from step 10 (or steps 8, 6, 3, 2, or 1 as needed).

13. Deliver the results in multiple formats, including documents and meetings.

14. Follow up as needed with clarification, presentations, and so forth.

We won't detail every one of those steps here, but we would add a few observations. First, it should be evident that the overall *timeline* for a project can vary tremendously based on the complexity of the various steps. There are three factors that often slow down Quant UX projects:

- Stakeholders may be frustratingly vague on the questions and not show enthusiasm for any of the possible directions (Steps 1 and 3).

- Data may be difficult or impossible to obtain, or they may turn out to be unsuitable due to collection errors, noise, or format (Steps 2, 6, and 7).

- The statistical models may fail, break, or prove unsuitable (Step 8). We recommend that you plan ample time and set expectations to allow for iteration at any step.

Second, you may notice that most of the steps are not specifically *quant* steps in any hard sense. That is, they don't involve coding, working with a data set, or building statistical models (which are Steps 6-9). Most of your time and mental energy may be spent clarifying problems, finding data sources, keeping stakeholders informed, creating presentations or whitepapers, presenting results, and answering questions (see Chapter 13). If you expect that a Quant UX role will be mostly devoted to a narrow form of "working with data," this may be surprising or disappointing.

Third, you can maximize the odds of success—in obtaining data, performing analyses that work, and finding something suitable to deliver to stakeholders—if you adopt a multidimensional, multimethod approach. You should try to answer a question not with a single outcome metric but with more than one metric (thus *multidimensional*; see Chapter 7). Do not rely on a single method either of data collection or analysis (thus *multimethod*). No single metric or statistical method can give a complete picture of the complexity of users. By using a few metrics and methods in combination you will gain confidence in results and be prepared to answer a broader set of questions.

4.7 Key Points

In this chapter, we emphasized that Quant UX research is part of UX research. The most important aspect of understanding UX research is to focus on research problems from a user-centric perspective (Section 4.2).

Beyond that, key points to emphasize are as follows:

- Quant UX exists at the intersection of technical skills in statistics, programming, and human-centered research (Section 4.1).

- Quant UX researchers have a T-shaped skills profile. This includes breadth and basic skills across the Quant UX space plus depth in at least one skill area (Section 4.1.1).

- We outlined five principles for user-centric research: adopt the user's perspective; assess user-centric variables and outcomes; answer "why?" with a cognitive approach; focus on unmet needs; and relate to UX actions and stakeholders (Section 4.2).

- Industry research too often neglects basic questions of research validity (Section 4.3). Always consider whether a method or approach is logical and grounded in realistic expectations.

- Often UXRs are assessing users, in which case it is important to gather large, representative samples. At other times, however, UXRs may be assessing a product, in which case it is significant to find even a single occurrence of an important problem. Our research planning, analyses, and reporting should take account of the difference (Section 4.4).

- The ethics of conducting Quant UX research should not be assumed or taken for granted (Section 4.5). The legal requirements for conducting research and preserving privacy are a minimum bar, not a definition of what is ethical.

- It is easy for UX and Quant UX researchers to slip into reporting results that are shaped to be "useful." You should always consider whether you are reporting what is most important from the *user's* perspective. Focusing narrowly on a team's goals may lose sight of what users are saying (Section 4.4).

- Research plans are usually complex and involve many more steps than simply collecting and analyzing data (Section 4.6). Allow time to plan, develop methods, find data, and iterate.

- Many Quant UXRs spend far more time on working with stakeholders, finding data, pursuing methods that don't work out, and communicating with teams than they do in directly working with data sets and statistical analyses (Section 4.6).

4.8 Learning More

The issues in this chapter are crucially important to the "UX" part of Quant UX and each one of them could be explored in much greater detail. Following are some pointers to get started in each major area.

Thinking like a UX researcher. Don Norman's book *The Design of Everyday Things* [101] is a classic and enjoyable introduction to thinking like a UX researcher or designer. David Travis and Philip Hodgson give a good account of how UXRs approach research—although with a primarily *qualitative* emphasis, not quant—in *Think Like a UX Researcher* [141]. If you are interested to try traditional qualitative research alongside Quant UX, Jakob Nielsen's *Usability Engineering* [99] is admirably compact, readable, and dense with insight.

Quant UX research methods. Apart from this book, there are no books that are highly specific to Quant UX as we define the field. The closest previous text has been Sauro and Lewis, *Quantifying the User Experience* [120], although it is targeted primarily to qualitative researchers who are interested in basic statistics. A general paper by Chris and colleagues presents a high-level overview of Quant UX research methods [28]. Learning generally about statistical methods is also helpful, of course (see Chapter 5), although those are not specific to Quant UX. The Google Research archive for Human-Computer Interaction and Visualization may be the largest single set of papers related to Quant UX (along with general UX). At the time of writing, it is available at `https://research.google/research-areas/human-computer-interaction-and-visualization/`.

Research validity. Many texts have been written about research validity and reliability; if you have taken graduate-level courses in research design or statistics, you may have a favorite. One of our favorites is Maxwell, Delaney, and Kelley, *Designing Experiments and Analyzing Data: A Model Comparison Perspective* [91]. Although it is dressed up as a statistics text (and a good, readable, and interesting one), we appreciate it especially for its strong emphasis on research philosophy. It demonstrates the importance of careful thinking about research design and not simply applying formulaic methods. Also, thanks to its grounding in psychology research, it is especially close to the kinds of human-centered questions and data questions that arise in UX research.

Research ethics. The term "ethics" is overloaded. Depending on the context, it may refer to institutional procedures (such as the "ethics code" of psychologists or a given organization's policies), to public policy and social implications, or to philosophical theories and analyses. When you get into a conversation about ethics, you should clarify which of those areas is being addressed.

A concise overview of general procedures and ethical frameworks applied to data science is Martens, *Data Science Ethics* [89]. A collection of short articles, similar to opinion editorials, addressing an assortment of topics is *97 Things about Ethics Everyone in Data Science Should Know,* edited by Franks [45]. A forceful examination of the social implications of data science and algorithmic fairness is Brandeis Hill Marshall's *Data Conscience: Algorithmic Siege on our Humanity* [88].

Brian Christian's *The Alignment Problem* [32] discusses the difficulty of building machine learning systems that comport with practical and ethical expectations. We highly recommend it as an overview of the general problem of aligning empirical models with ethical expectations and for its readable accounts of important cases where models failed, to their creators' surprise.

For philosophical issues relevant to UX research and design, Chris has written about the relationship between design and ethical construction [19], and suggested ways in which UX researchers may productively engage users directly in such considerations [18].

An example of a comprehensive professional code of ethics is the American Psychological Association's *Ethical Principles of Psychologists and Code of Conduct* [2], which is effectively written into law for professional psychologists by regulations in some US states. Although the APA ethics code is specific to the practice of psychology, it is worth reading for its general framework and considerations that apply to all human subjects research.

## CHAPTER 5

Statistics

The most crucial skill for Quant UXRs is working with data and drawing inferences from it—that is to say, *statistics*. As noted in Chapter 3, "Quantitative UX Research: Overview," Quant UX requires fluency with statistical models at a foundational level, not necessarily at an advanced level.

In this chapter, we describe the level of statistical fluency that is required for Quant UX research, including both the minimum requirements and skills that demonstrate higher levels of expertise. We also describe the importance of statistics over machine learning, and review several common misunderstandings we have observed in the practice of statistics.

In addition this chapter's "Key Points" section (Section 5.5), we summarize our recommended knowledge of statistics in the example hiring rubrics in Appendix B.

5.1 Why Statistics?

The fundamental goal of UX research is to learn about users and their behavior, and to use that insight to draw inferences that inform product development. A more subtle point is that such inferences are always uncertain and depend on the quality of our data, our samples, and our models. UXRs try to get high-quality and representative data, reduce uncertainty, and appropriately communicate uncertainty to other stakeholders. The discipline of *statistics* is concerned with all of those issues.

5.1.1 Statistics vs. Machine Learning

There is huge overlap between *statistics* and *machine learning* as well as much confusion about the differences. The statistician Leo Breiman characterized them as "two cultures" in a well-known paper [10], describing machine learning (ML) as being most concerned with black box *prediction*, and statistics as most concerned with *modeling*. We differ with

© Chris Chapman and Kerry Rodden 2023
C. Chapman and K. Rodden, *Quantitative User Experience Research*,
https://doi.org/10.1007/978-1-4842-9268-6_5

Breiman on a variety of details (for reasons similar to those given by Gelman [49]) and won't pretend to clear up all of the issues. We believe, however, that Quant UXRs are most naturally aligned with the statistics side of the divide.

Statistical modeling involves a researcher's conception of an implicit *data generating process* (DGP). For example, consider a respondent's answer to a survey question about their satisfaction with a product. This can be understood as reflecting some underlying *latent* real satisfaction, along with systematic deviation due to scale or response bias, plus random error due to respondent inconsistency and other factors. Each of those— the observed response, the presumed latent score, the systematic deviation, and the random error—might be specified and estimated in a statistical model. By contrast, an ML approach might simply optimize for observed accuracy of prediction, and not be concerned with the presumed structure of the data.

The statistical approach naturally aligns with UX research for the simple reason that the core research goal of UXRs is to understand users. That is not to say that ML approaches are wrong or useless. We are in favor of using any valid approach that is helpful. Yet we believe that approaching questions from the statistical perspective— focusing on the presumed mechanisms of users' responses, behaviors, and associated sources of error and uncertainty—is ultimately the more helpful of the two approaches for Quant UX goals.

5.2 The Foundation: Sampling and Data Quality

Nothing is more important for analytics than good data. It is unfortunately far too common that analysts use low-quality data, often because they have no choice. Management or stakeholders may present an analyst with poor data and insist that it be used.

Another common pattern is that an analyst knows that a data set is suspect but decides to "see what I can find." The problem with this is that you can almost always find a story that sounds compelling, regardless of data quality. Yet with poor data, that story is likely to be false (see Section 4.3).

Unfortunately, there is no simple test for good data or bad data; it requires domain knowledge, understanding of statistics, and interpretation. However, we can identify some factors to evaluate with regard to data quality.

First, were the data collected from a representative sample? A common issue here involves *samples of convenience*, perhaps arising from voluntary participation in research by opting in or responding to a survey. A sample of convenience also arises when data collection occurs through a unique channel such as a beta program or event. Those who participate may differ systematically from those who do not. In one case, Chris observed two satisfaction data sets for the same product. One was collected via a survey at an in-person event while the other was obtained via an email survey to all of the product's registered users. Among customers at the event, product satisfaction was over 80%. That's great, right? Unfortunately, when the larger set of users was surveyed by email, they reported *dissatisfaction* greater than 70%. That is a net difference of at least 50% of surveyed users, highlighting the importance of sample quality and representation.

This does not imply that samples of convenience are useless. Rather, you should use them cautiously and compare with other indicators. Do not compare convenience samples to any absolute standard (such as aiming for "90% satisfied"). Instead, compare them to one another *relatively* over time to see whether an outcome changes, apart from the question of full representation. (We present another example, about satisfaction ratings, in Chapter 8, "Customer Satisfaction Surveys.")

Closely related to the sample source is whether we may assume *random sampling* and *lack of confounding factors* in respect to a question of interest. Suppose that we want to compare customers to non-customers on some measure of interest, and we send them a survey. If we compare the two samples with a statistical model such as a t-test, there is an implicit assumption that the samples were randomly drawn and that any confounding issues are similar for the two groups. Yet that is unlikely to be the case for these samples; customers are usually more likely to respond than non-customers, and should be expected to differ on many dimensions.

With online sampling, a very common confound is to have selection odds that are proportional to usage frequency. For example, suppose we want to survey 1000 users and do this via a random intercept survey that is triggered for 1% of all users until we get 1000 responses. Is that a random sample? The question is, "A random sample of *what*?" If the answer is "a random sample of users" then the answer is most definitely *no*, it is not a random sample. Why not? Because our most frequent users will have vastly greater odds of being sampled. That may be adequate for some research questions—for example, if we want to collect feedback from frequent users—but it is not a good sample of *users* overall.

Another set of issues involves the *validity* and *reliability* of the data, as previously introduced in Section 4.3:

- *Validity*: Do the data truly mean what we think they mean? Consider the ubiquitous question, "How likely are you to recommend ___?" There is an underlying assumption that the respondent will even recommend the product at all. Many common products are very rarely discussed or recommended (see Section 4.2.3). In the case of a survey, pre-test it with actual users to determine how they interpret the items.

- *Reliability*: If a data point were acquired twice, would the values agree? A good practice is to use multiple indicators, ideally obtained through multiple sources.

An *outlier* is a data point that is outside the expected range for a statistical model. Generally speaking, our view is that all data points are good data points, as long as they come from an accurate and intended sampling process. Under that view, we prefer not to remove outliers but rather to consider whether they will have an unintended or disproportionate effect on a model; robust statistical models are often useful. One thing we never recommend is to remove data simply on the basis of a preconception that extreme values should be trimmed. Rather, we would explore the origins of such data and make an informed decision. We usually opt to retain data rather than delete it, when there is any question.

Finally, you must be diligent to check the accuracy and reliability of *data pipelines*. Are there technical problems leading to loss or inaccuracy of data? In any data cleaning process, consider formal tests of the degree of change or deletion of data.

A data set is never perfect, and it would be unreasonable to object to learning from data just because a particular data set has flaws. Instead, a researcher should work to collect good data, carefully consider and communicate limitations, and compare and contrast results with those from other data sets.

5.3 Core Statistical Analysis Skills

For Quant UX work, a researcher should be fluent in basic statistics. In our view there are five crucial areas where Quant UXRs should have relatively strong knowledge and experience. In priority order, Quant UXRs should have expertise in

1. Exploratory data analysis and visualization

2. Descriptive statistics

3. Inferential tests and practical significance

4. Fundamentals of A/B testing

5. Linear models (linear regression)

Among those items, each is well covered by courses and textbooks in statistics, except the latter part of the third item, "practical significance," which is difficult to understand from a textbook. Ability to interpret the practical implications of a statistical result is a skill that develops over the course of working with real data sets and explaining analyses to others. We hope our examples in Part III will help build your understanding.

Before diving into the five areas, we also want to debunk the common misconception that "statistics" is a set of *procedures* to apply to data. We often see courses and textbooks promoting that view, with prescriptions such as, "For *these* kinds of data, apply *this* statistical test." Such recipes are often substantively incorrect, for example when they suggest that scalar properties are the most important aspect of data points (such as whether a variable is ordinal or continuous). Even more worrisome is the underlying, premature assumption: that we are undertaking a well-defined and presupposed task of data description, reduction, or inferential testing.

A better way to consider statistical analysis is that it involves *learning from data*. It is only after understanding your data in detail that you should proceed to such steps as summarizing it, applying statistical tests, or otherwise modeling it.

For readers who are steeped in traditional academic approaches to statistics, this approach is quite different from *hypothesis testing*. Quant UXRs often use data to learn about user behavior with no predetermined hypothesis. That doesn't mean that inferential testing is unimportant but rather that it is only one approach among several that are important in Quant UXR practice.

5.3.1 Exploratory Data Analysis and Visualization

The first step in learning from data is exploratory data analysis (EDA). In Section 5.2 we discussed preliminary issues to consider, to help ensure that you have reasonable data. In this section, we assume that you have a data set that meets these basic requirements, and now you want to start learning from it.

The term "exploratory" is the crucial concept here, as it implies that you should approach data with few preconceptions and look at what it tells you. As with many topics in this book, EDA is a large topic with a substantial literature. Following are some general practices to consider as a minimum starting point:

- Examine the range (minimum and maximum) and quantiles of all numeric data points. Do they make sense?

- If you have text data, read some of it. For surveys with open-end comments, read them. Are there any problems?

- Make individual scatter plots, density plots, and frequency plots, as appropriate, for all important numerical and categorical variables.

- How much missing data do you have? Is that a reasonable amount?

- Select a few complete observations and examine them in detail. Are the observations realistic?

- Did you clean any data, or remove outliers? Why? How would you know whether that improved data quality as opposed to applying opinions that biased it?

- Plot many (or ideally all) pairs of variables against one another. A *scatterplot matrix* is one way to do this (see Section 4.4 in [25]).

Finish those steps before performing further analysis such as summarizing the data in terms of central tendency (mean, median) or fitting statistical models. For some kinds of data and research problems, further exploration may also be warranted, such as correlation analysis or exploratory factor analysis.

What about *data visualization*? Our opinion is that it is not a separate skill but should be a part of almost every step in data analysis. When you get data, plot it. Data summaries such as the mean and distribution? Plot them! A statistical model? Plot it. Results? Plot them! Takeaways for stakeholders? You guessed it: plot them.

Although there are specialized and quite useful data visualization methods for particular kinds of data, such as the sequences sunburst (see Chapter 9, "Log Sequence Visualization"), we emphasize that data visualization should not be seen as an optional, fancy, difficult, or unique skill. Rather, it is a core skill of all statistical analysis because it is essential to developing an understanding of your data.

5.3.2 Descriptive Statistics

The second important area of statistical fluency is *descriptive statistics*, describing a data set in shorthand ways. When you hear about a mean, median, or standard deviation, those are descriptive statistics.

In this area, there are four crucial aspects to consider:

1. Which summary statistics are appropriate for the data at hand? For continuous data, you might consider forms of arithmetic central tendency and range, whereas ordinal or nominal data might be summarized by frequencies. This consideration is taught in every basic statistics course.

2. It may be helpful to transform your data in some way, especially if it is typical of a skewed distribution. User data often contains values that are typical of power law or logarithmic distributions, comprising many small values with a long tail of infrequent large values. Household income, number of visits to a website, productivity per developer, willingness to pay for a product, number of soda cans in the back seat of a car...all are examples of such data. Consider transforming skewed data using a logarithm, inverse value, or other mathematical function as appropriate (there is a list in Section 4.5.4 of the R companion [25]).

3. A less obvious question is, "What are the particular points of interest to describe the data?" Analysts often unthinkingly report the *mean* and *standard deviation*. But why are those the best values? For skewed data, the median may be better; or for a given business problem, you might be more interested in the top 10% of customers, or those who fall below some cut-off point. For instance, dissatisfied customers may have much to tell you, and you may want to know how many there are. It is important to include both specific points—such as the mean, median, or some percentile—as well as some characterization of the distribution (density, variance, and so forth).

4. Closely related to the previous points, consider *robust statistics* that do not change when a small proportion of extreme values change. When Chris worked at Microsoft, a running joke was that various colleagues had been "billionaires on average"...namely, when they met with Bill Gates.

    Of course their wealth did not change when Bill Gates entered the room, but the *mean* value of the set of people did change. A better statistic in such cases would be the median or a set of percentiles.

As noted in the previous section, statistical skills are incomplete without considering appropriate data visualization. For descriptive statistics, box-and-whisker, density, frequency, and violin plots are good candidates for visualization during exploratory data analysis. However, we would be cautious about sharing them with stakeholders who are not deeply knowledgeable about statistics (see Chapter 13, "Research Processes, Reporting, and Stakeholders").

Surprisingly often, a Quant UX project needs no statistical analysis beyond descriptive statistics. For example, consider an e-commerce website. If a team wants to know how satisfied its users are, how much they spend, how many new users the product has gained, how often existing users return to the site, and what percentage of them successfully get through account signup, all of those might be answered by descriptive statistics. (Those questions also map to the categories of *happiness* (H), *engagement* (E), *adoption* (A), *retention* (R), and *task success* (T) in the HEART framework, as we discuss in Chapter 7.) Before engaging in advanced modeling, a Quant UX team should have a strong understanding of basic user data.

5.3.3 Inferential Tests and Practical Significance

Inferential testing, often called *statistical testing* or *significance testing*, looks at the relationship between two (or more) variables. It asks whether an observed relationship—such as a difference in site visits, according to customer type—is likely to be a real difference as opposed to a random relationship. When we say that it "asks," we are being intentional; our view is that inferential testing *asks* the question about real differences yet rarely gives a definitive answer on its own.

The problem with inferential testing of differences and reporting of statistical significance is that what we really want to know is whether a difference is *meaningful* to the product team, user, or business. In the most common forms, statistical significance reflects a composite metric based on average difference, sample variance, and sample size. Of those three, the average difference and observed variance may be important for our decisions, but the sample size in itself is of little interest; a large sample may be hopelessly biased, while a small one may be informative. As we noted in Section 5.2, the more important question about a sample is whether it reflects the population of interest.

Does that mean that statistical significance is unimportant? No. Statistical significance is an important and yet ultimately small aspect of whether an analysis is useful. (As an aside, this includes Bayesian variations of model comparison using credible or compatible intervals; see Sections 6.2 and 6.6 in the R companion for discussion of Bayesian approaches [25].) When it applies, we would say that statistical significance is a necessary condition for a finding to be reported but it is not sufficient; an effect should also be large enough to be meaningful for the user or product. More often, we suggest that statistical significance should not be reported. An analyst might or might not consider it, according to the context, but it is not a relevant topic for stakeholders and partners. Effect sizes and confidence intervals are more interesting and useful.

How large of an effect is large enough? It depends! Consider an A/B test (discussed in the next section) where we see a 1% difference between designs and, based on thousands of users, that difference is statistically significant. Is the 1% difference important? If it relates directly to millions of users or millions of dollars—where 1% of 10M = 100,000 users or $100,000—then it may be sufficient to make a decision. On the other hand, if the effect is nebulous or reflects an indirect influence, we would look for a much larger difference or other corroborating evidence.

As a Quant UXR, you should be familiar with the following concepts for inferential testing:

- Standard deviation in a sample

- Standard error of the mean (of a sample or group)

- Chi-square and other methods to assess frequencies and nominal data

- Standard error of test statistics such as a correlation coefficient

- T-tests and when they are appropriate

- Linear models and their evaluation (see Section 5.3.5)

As with every skill in this book, there is much more that you might learn, according to your interest or the needs of a particular organization. Two particular areas are bootstrapping methods and Bayesian alternatives to statistical significance. Section 5.6 suggests paths to learn more.

5.3.4 Fundamentals of A/B Testing

An *A/B test* is a randomized experiment where two or more versions of a design or other intervention are compared side by side for their effect on some outcome of interest. For example, we might test two designs of a sign-up process (the intervention) to see which one leads to a higher number of new users (the outcome of interest). Or we might test multiple sizes or colors of a button, to see which one gains more clicks or conversions. One condition, usually the current design, is the *control*, and the other conditions are known as *treatments* or *variations*.

Such tests provide teams with the ability to make changes to the design of a product while being confident that those changes are leading to measurably better outcomes. This is extremely useful for teams because it is surprisingly hard to predict whether a given change will be an improvement, even for the most experienced practitioners.

The analysis of an A/B test is a special case of inferential testing (as previously noted); the goal is to determine the statistical and practical effect of a predictor variable (the design intervention) on a dependent variable (the outcome of interest). However, it is easy to over-focus on the *test* aspect of inferential testing and not pay enough attention to the potential *inference*. If the experiment is set up well, then the statistical analysis itself may be straightforward, such as a t-test, or more complex, such as a matched sampling model. The analysis may even be performed automatically by internal tools or third-party A/B testing software. The more important question, however, is how well the experiment was set up to focus on the inferential question at hand.

What do we mean by focusing on the inference? The most important aspect of an A/B test is the experimental design—exactly what are you testing, how, and with whom? Quant UXRs are often asked to help design these experiments because they are members of the UX team. For example, a UX designer may ask a Quant UXR to set up an A/B test to compare designs, or a PM may ask a Quant UXR whether it is possible to use an A/B test to compare two features. To do this, the Quant UXR will need to consider many questions for the experimental design. Those include the following:

- Exactly who is the user of the product?

- How can we sample those users effectively?

- What is an appropriate format for the A/B intervention?

- Are there confounding and possibly unobserved variables that pose problems for analysis?

- How will this outcome metric relate to other outcome metrics?

In many cases, those questions are surprisingly complex to answer. Here's a contextual example. Suppose we have a free streaming video service and want to gain conversions to a paid subscription service. Our designers have created two interventions to invite sign-ups: a sidebar invitation that users can click at any time, and a pop-up window that requires a response ("sign up" or "not now") before they can continue. From one perspective, the question is easy: show each design to some random viewers and look at the resulting proportion of subscriptions in each group (this might be a linear model with a binomial treatment such as a t-test, or a multinomial treatment such as ANOVA).

Yet before arriving at that point, a Quant UXR should consider some of the following points. First, who is a "user?" Does it mean a *person*? Or a streaming *device*? Or a signed-in *account* with a unique identifier? Or a paying *customer*? Depending on the product and design, the differences may be profound. Second, will it be possible to sample from that population? Sampling accounts, for instance, does not sample people; and sampling devices does not sample accounts. Third, how does the sampling interact with other behaviors that may pose questions for our interpretation of results?

In the video subscription case, if we are interested in paying subscribers, then we are likely most interested in sampling *accounts*, under the view that an account is more easily sampled than a person and is usually also the level of payment (a customer). And it abstracts away from the differences among devices. However, that implies that we will need to consider how the presentations of the A and B conditions may vary across devices.

A common issue with online sampling is that the odds of selection are often proportional to usage frequency (see Section 5.2). In an A/B test, it may be the case that one of the conditions has a closer relationship to frequency of usage than the other, making it difficult to meaningfully say that one design is better. In the streaming video service example, let's assume that equal numbers of accounts are selected into the sidebar and pop-up conditions. Some of those accounts will have more site usage than others. In the sidebar condition, more frequent users may be somewhat more likely to notice the sidebar. Since the pop-up is more obtrusive, more frequent users will encounter it more often (assuming it is triggered at regular intervals), and therefore may be disproportionately affected by this condition. It may make more sense in this situation to use qualitative research methods to help choose one of the designs, and then use A/B testing to optimize that design.

Finally, it is extremely important to choose the right outcome metrics (ensuring that the tests are measuring improvements in user experience), and avoid excessive optimization of a single metric. If the pop-up design leads to more sign-ups but also more user annoyance or higher abandonment of the product, is it worth it? We discuss these issues further in Chapter 7.

We can't elaborate on every aspect of A/B testing here, and we point to additional resources in the "Learning More" section (Section 5.6). Meanwhile, we emphasize this general point: *the most important aspects of statistical analysis occur before any data are collected*, when you decide on the details of the experimental design. Whether Quant UXRs are called upon to perform the final statistical analysis or not, they are expected to think about how a team can field experiments appropriately to answer business questions about design and feature options.

5.3.5 Linear Models

The final piece of a Quant UX statistics foundation is fluency in *linear models*. By linear model, we mean the general family of statistical methods that estimate the degree to which one or more predictor variables influence a continuous or quasi-continuous outcome variable. Linear regression models are the most common variety and the most essential to understand.

Unfortunately, the simplicity and availability of tools for linear regression—such as statistics add-ons to spreadsheet applications and point-and-click statistics packages— have led many analysts and stakeholders to overestimate their knowledge of such models. The ability to fit *some* model does not indicate understanding, appropriateness, or ability to interpret results.

As a Quant UXR, you should have a good working understanding of the following issues in linear modeling, in either a classical ("frequentist") or Bayesian perspective:

- The kinds of data that are appropriate as predictors and outcome variables in linear models

- The effects of multicollinearity (correlation among variables) and how to reduce its impact

- Transforming data for better interpretability or to standardize it (zero-center, and transform to units of standard deviation)

- Choosing an appropriate model for the variables, especially the outcome variable (for instance, understanding when you should use a linear regression model as opposed to some other kind of model or analysis)

- Making a model robust in the presence of skewed data or extreme values

- Deciding whether to include interaction effects as opposed to only main effects

- Fitting the model and assessing goodness of fit, including residuals

- Comparing models to assess comparative fit and, more importantly, suitability to answer the research question at hand

- Interpreting standardized and unstandardized coefficients as appropriate for a given business question

- Reporting confidence intervals and interpreting standard errors for estimated values

- Visualizing any results

The crucial aspect is to understand the practical importance of those elements. Knowing a set of formulas is unnecessary; rather, you should have hands-on experience and be able to explain the advantages and disadvantages of various modeling choices. A hands-on example of approaching linear modeling in this way is given in Section 7.3 in the companion R [25] and Python [127] texts.

Linear modeling topics that are not required (although useful) include more complex multivariable models, multinomial models, structural equation models, multilevel and mixed effects models, graph and causal models, factor analysis and principal component analysis, and latent class models. We recommend not to be concerned with those unless statistics is your area of in-depth expertise.

5.4 Frequently Observed Issues

We want to emphasize a few problems that we observe repeatedly. As a Quant UXR, you may expect each one of these to appear often in your work with stakeholders, other analysts, and in job interviews. Some of these have been mentioned already in this and other chapters, yet we want to reinforce them and compile them in one place. Although none of these issues has a perfect solution, each deserves awareness, consideration, and potential mitigation.

5.4.1 Bad or Biased Data

As described in Section 5.2, analysts too often work with data sets that are deficient. The data points may be systematically biased due to the acquisition method or technical flaws; errors may be introduced in the process of data cleaning; and the data points in themselves simply may not mean what you *assume* they mean. There is a relevant, famous quote from the statistician John Tukey, "The combination of some data and an aching desire for an answer does not ensure that a reasonable answer can be extracted from a given body of data." [143].

5.4.2 Focusing on Discovery, Losing Sight of Decisions

Those of us who work with data often become caught up in the excitement of understanding users, and stakeholders often share the view that the goal of UX research is to understand users. We believe this is admirable but incorrect; our task is to help stakeholders to *make decisions* about products. Understanding users' needs, behaviors, and attitudes often assists us to make better decisions, but understanding in itself is not the goal. One might look at this in terms of expected utility: because the range of possible knowledge about users is approximately infinite, the expected utility of any particular knowledge—considered in isolation from a specific need—is near zero. We urge you to return over and over to the question, "What decision are we making?"

5.4.3 Prematurely Assuming an Outcome of Interest

The landscape of possible user data and product considerations is daunting. It would be wonderful to reduce it to a simple outcome metric or set of metrics. Unfortunately, we believe that is more likely to go wrong than to go well. People are complex, products are complex, the interaction of people and products is complex, and competitive responses from other products are complex. On the stakeholder side, there is often a desire to reduce assessment to a single "North Star" metric such as customer satisfaction, revenue, active user counts, or likelihood to recommend the product. This may be combined with a goal that is defined in advance of any data, such as "achieve 90% customer satisfaction."

On the analyst side, this problem may arise when the analyst makes a choice to emphasize something based on simplistic assumptions, such as, "Of course we care about *[some metric]* because *[some plausible reason]*." For example, "Of course we care about profit because every business must be concerned with profit," or "Of course we care about active user counts because active users drive everything else."

The problem is that the data are likely to be interconnected in complex ways and a single outcome cannot represent real-world complexity. As we already said, we recommend to focus on the decisions to be made and work backward from there to identify relevant data and models. In our experience, product decisions rarely involve a single outcome metric. We discuss this in depth in Chapter 7 and review additional examples in Chapter 8 and Chapter 9.

5.4.4 Interpreting Statistical Significance

When we decide to answer the question, "Is a result statistically significant?" we find that stakeholders and analysts often then want to know, "How significant is it?" This is positively encouraged by features of the inferential testing environment, such as the appearance that p-values are continuous, and the persistence of practices such as journals that require results to demonstrate "statistical significance, $p < .05$". Those are misinterpretations that you should avoid. Statisticians have warned of the misunderstandings and "fantasies" [14] of statistical significance for decades.

As we discussed in Section 5.3.3, you should give far more attention to questions of effect size and whether a finding will have practical significance for a decision to be made.

5.4.5 Applying Fancy Models

The final problem that we frequently observe is when an analyst uses a complex model but is uncertain what the model actually does or the trade-offs involved in applying it.

It is relatively easy to pick any one of thousands of statistical analysis procedures from libraries in R or Python, or menus in SPSS. Doesn't it sound great to "segment users with a hyperplane kernel model"? It is great—but only when it is needed.

Whenever you use a model, make sure you know what's really occurring, what the model's assumptions are, what alternatives you are choosing *not* to use, and how to interpret the results. Ideally, compare models and demonstrate to yourself (stakeholders won't care) that the model you choose is better than plausible alternatives.

Usually we find that descriptive statistics, data visualization, and relatively common models are able to answer most of the questions that confront us in Quant UX research.

5.5 Key Points

For Quant UXRs, it is far more important to be fluent in basic statistics than advanced modeling. We believe that most problems in UX data analysis are addressable through careful attention to data sources, exploratory analysis, and straightforward models. Focus on product and business *decisions* rather than on collecting general, unfocused information about users.

The following skills are fundamental for the practice of Quant UX research:

- Effective sampling design and attention to data quality (Section 5.2).

- Understanding UX data as coming from human data generating processes (DGPs) and the implications this has for noise, collinearity, and response bias (Section 5.1.1).

- Starting any analytic project with thorough exploratory data analysis (EDA) and data visualization (Section 5.3.1). Pay particular attention to data ranges, distribution, and assumptions in data cleaning.

- Using descriptive statistics appropriately (Section 5.3.2). Many questions can be answered through effective plotting of data with descriptive points and distributions.

- Inferential testing and understanding the crucial difference between statistical significance and practical significance (Section 5.3.3). Do not focus on statistical significance with stakeholders.

- Appropriate experiment design for A/B tests, to ensure that they answer the inferential questions, plus the relevant statistical approaches (Section 5.3.4).

- Linear modeling, especially a strong understanding of its assumptions, effective usage, and how to interpret the resulting estimates with regard to a business question (Section 5.3.5).

On the reverse side of those skills, there are several problems to avoid:

- Using bad or biased data and assuming without evidence that it is good or "good enough" (Section 5.4.1).

- Giving too much attention to "discovering" facts about users rather than addressing product decisions (Section 5.4.2).

- Assuming that some specified outcome metric—perhaps satisfaction, revenue, or usage—has a clear meaning and is an obvious primary goal. Most of the time, a product team should consider multiple metrics, where any single indicator is fallible (Section 5.4.3).

- Interpreting statistical significance as if it were an indicator of importance (Section 5.4.4).

- Using complex models, especially ones you don't understand, when a simpler model might be just as effective at answering a question (Section 5.4.5).

In Chapters 7–10 we demonstrate specific examples of research and analysis in line with these points.

5.6 Learning More

For many analysts, learning about statistical methods is a lifelong endeavor. The field of statistical methods is deep, varies across disciplines, and is continuing to grow with the development of new mathematical models, visualization techniques, computational tools, and ways to apply them. We hope that you will view the field's diversity and depth as inspiring and reassuring, rather than daunting.

Foundations. For readers who are relatively new to statistics, we recommend focusing on basic methods coupled with *hands-on application* to problems of real interest to you. There simply is no substitute for working with real data sets and paying attention to the ways in which your assumptions about data are challenged by real observation.

A/B testing is a good starting point to learn about statistics because it is a very common area of analysis in UX research. To understand more about how A/B testing is used in industry, we suggest two books from authors with many years of collective experience working at large tech companies. Kohavi, Tang, and Xu's *Trustworthy Online Controlled Experiments* [72] is a comprehensive and practical guide to the topic, while King, Churchill, and Tan, *Designing with Data* [69], is focused on the application of A/B testing to design and UX. The Python [127] and R [25] companion books cover the most common statistics needed for A/B experiments in Sections 5.2 and 6.5 in each book. McCullough's *Business Experiments with R* [92] is a complete presentation of the details of experimental design and statistical analysis for applied experiments.

In addition to A/B testing, as a Quant UXR you'll want to build moderate depth in general statistical methods, as follows. This book's companion R [25] and Python [127] texts adopt a hands-on approach where you work first with provided data and then are encouraged to analyze your own data sets. Ismay and Kim's *Statistical Inference via Data Science* [62] will also give you hands-on experience with real data in R. Its topics closely parallel our recommended skills in this chapter.

If you don't want to program in R or Python (or a similar language)…well, first we'd try to convince you that programming is a good thing (Chapter 6, "Programming"). Beyond that, if you are not ready to switch completely to a programming environment, we encourage you to consider using a modern data analysis suite such as JASP [64]. An excellent text that covers every skill in this chapter is *Learning Statistics with JASP* [96]. JASP is built on top of R, so you can learn statistics in a GUI environment and then consider later moving into the complexities of the R language.

Beyond the basics. Among more advanced statistical analysis skills, we suggest five areas that are especially useful in Quant UX: psychometrics, structural models, Bayesian data analysis, causal models, and clustering and classification. Note that *none of these advanced topics is required* for Quant UXRs. We suggest them here with regard to the T-shape distribution of skills (see Section 4.1.1). They are areas where someone strongly interested in statistics could consider developing personal depth, because each of them can be very valuable for specific Quant UX projects.

- *Psychometrics*: Focuses on the data-generation aspect of human data. When we collect user data—especially self-report or survey data—is it reliable? Does it mean what we think? How does it relate to other measures? Psychometrics helps us answer all of those questions. A comprehensive yet approachable guide is Furr, *Psychometrics: An Introduction* [46].

- *Structural models*: Relate several or many observations to one another in a more complex way than a linear model. For example, consider the concept of a user's likelihood to recommend a product. That may be influenced by satisfaction and by the opportunity to recommend. Satisfaction will be influenced by the user's behaviors with the product, needs in their environment, and expectations. The opportunity to recommend may itself be influenced by the same set of needs in their environment as well as other factors such as location and status. Given appropriate data, a structural model (such as a *structural equation model*, SEM, or a *directed acyclic graph*, DAG) may allow one to model these interactions and assess the influence of each variable on the others. The R companion text introduces structural models in Section 10.1 [25]. Kline's *Principles and Practice of Structural Equation Modeling* [70] is a definitive guide with attention to best practices.

- *Bayesian data analysis*: Bayesian methods offer several advantages over classical statistical procedures, especially for eliminating the traditional concepts of "significance" and allowing estimation of models in situations with complex relationships and sparse data. The Python [127] and R [25] companion texts introduce Bayesian models alongside traditional models. For a complete, enjoyable, eye-opening, readable, and all-around excellent grounding in Bayesian statistical methods, we highly recommend Richard McElreath's *Statistical Rethinking* [94].

- *Causal models*: Statistical models are often misinterpreted as being about *causation*, and we assume you are aware that this is generally untrue. However, in recent years, there has been extensive work to go beyond the simple truism that "correlation is not causation." Can statistics help us to assess causation beyond simple experiments? The answer is *yes*—sometimes—and there is a great amount of work to develop new statistical approaches to assessing causation. Causal models build on the foundations of structural models noted above. Scott Cunningham provides a superb and fun introduction in *Causal Inference: The Mixtape* [38].

- *Clustering and classification*: Quant UXRs often get involved in clustering data, especially to find user *segments*. We have reservations about many of those projects and whether they are actionable for product decisions, yet clustering methods are a core element in intermediate fluency in statistics. Given some kind of group assignment—whether a user segment or some other outcome of interest—classification methods assign new observations to groups. For starters there is coverage of clustering and classification in the R [25] and Python [127] companion texts. For further depth, clustering methods are covered in depth by Everitt *et al.* in *Cluster Analysis* [43]. Outstanding coverage of classification methods using R is given in Kuhn and Johnson, *Applied Predictive Modeling* [76].

Finally, as we noted in Chapter 4, "UX Research," anyone who is building statistical or machine learning models should be familiar with the cases, challenges, and ethical concerns described in Christian's *The Alignment Problem* [32].

5.7 Questions and an Exercise

The statistics texts mentioned in Section 5.6 will be your best source for exercises in performing statistical analysis. In this section, we pose questions—as might occur during an interview—followed by an exercise in exploratory data analysis and basic statistics.

For the following questions, imagine that a product manager (PM) presents you with a data set that includes observations of 50,000 users' behaviors for 20 product features in a mobile application. A subset of 2000 users have given a rating of their satisfaction with the product on a 5-point ordinal scale. Among the 20 feature usage observations, 18 of the variables are binary (usage of the feature or not) while 2 are continuous (number of sessions and total time in the application by the user during the observation). Finally, there is a variable with assignment of users into three groups labeled "casual" users, "professional" users, and "disinterested" users.

1. What else would you want to know about the data set?

2. What steps would you take to determine the quality of the data?

3. How would you determine whether the behavioral observations
    appear to be reasonable? (Hint: this may include interaction with
    the PM as well as statistical analysis.)

4. How would you determine whether the satisfaction and time
    observations are reasonable?

5. How would you investigate collinearity (correlation) in the data
    set? If there was high collinearity, what would you do about that, if
    anything?

6. The PM wants to know which variables predict customer
    satisfaction. How would you go about modeling this? How would
    you know whether the model fits the data well?

7. The PM asks which features are more important to the casual
    users vs. professional users. How would you answer that question?
    (Hint: what does *important* mean?)

The following exercise is presented in several steps. The complete sequence applies several crucial methods described in this chapter. If it takes you more than an hour or so, we suggest that you work through one of the hands-on statistics texts suggested in Section 5.6. However, this example is not a complete test of the skills needed for Quant UXR and we encourage you to review exercises in the other recommended texts.

1. Load the data sets `statistics-dat1.csv` and `statistics-dat2.csv` from the book's website at `https://quantuxbook.com/data`. Did the data load correctly? How do you know for sure?

2. In each data set there are two variables, `v1` and `v2`. As far as you can tell from statistical analysis, are they good observations? Be sure to plot them.

3. In each data set, do `v1` and `v2` come from a similar data generation process? How do you know? (Note: this question is based on your *interpretation* of statistical analysis, not necessarily on a specific statistical test or procedure.)

4. In each data set, is there a statistically significant difference between v1 and v2?

5. In each data set, how well can you predict the values of v2 from v1 using a linear model? What would you say about that to a stakeholder?

## CHAPTER 6

Programming

The ability to program is a crucial skill for Quant UXRs. This may seem obvious or it may surprise you. In this chapter, we discuss why programming is important and describe which aspects of programming are essential for Quant UXRs.

Becoming a computer scientist is not necessary; the goal for Quant UXRs is to program *well enough*. As with every skill in this book, if you want to become a specialist, that is great, too. In this chapter we simply describe the foundation.

We don't attempt to *teach* programming here, because that is well covered in other texts. Rather, we wish to motivate those who are unsure about the relevance of programming to Quant UX or their own skills in programming, and set reasonable expectations for practitioners, managers, interviewers, and job candidates.

For a brief list of *hiring criteria* for programming skills, see the "Key Points" section in this chapter (Section 6.5) and the suggested rubrics in Appendix B.

6.1 Overview

Why is programming a crucial skill for Quant UXRs? Why is it insufficient to obtain data and use a graphical analysis package such as SPSS Statistics or Microsoft Excel to perform statistical analysis?

With the ability to program you will have the following:

- Vastly expanded abilities to obtain, process, and shape data for analysis

- Access to the widest range of tools and methods for statistical analysis and machine learning

- Improved accuracy of your analyses, thanks to the ease of repeating and improving them

© Chris Chapman and Kerry Rodden 2023
C. Chapman and K. Rodden, *Quantitative User Experience Research*,
https://doi.org/10.1007/978-1-4842-9268-6_6

- Greater efficiency of your work as you build a reusable, personal code library of methods and examples

- Opportunity to use and adapt code that others share online

- Ability to tailor data visualization for deeper exploration and more effective research presentations

- Enhanced collaboration with colleagues, as you share code with them

Those are impressive benefits, yet they come with some cost. First, if you are new to it, learning to program requires a great deal of motivated *practice*. Practice is essential because—just like a spoken, human language—the patterns must become imprinted through repetition. And it should be *motivated* by something of personal interest, because that will help you to persist and develop the patterns. Choose a project of interest, either at work or outside of work, and persist until it is completed. Then do another one!

The second cost is subjective: some people simply do not enjoy programming. It is a task that requires intense attention to detail along with highly structured thinking, which may relate to personality traits such as conscientiousness [66]. If you have previously tried programming, with a well-motivated project, and didn't like it, then Quant UX research may not be enjoyable for you.

6.1.1 Is Programming Required?

We are often asked by prospective Quant UXRs and by UX hiring managers whether programming is truly *required* as opposed to being merely *recommended*. The answer depends on the organization and details of the role. At some companies, programming is a formal requirement, and you may expect to undergo an interview to write code or *pseudocode* (text that is organized like code, with logic and formal structures, yet without concern for exact syntax, function names, and the like). At other firms, Quant UXRs are expected to focus on data collection or statistical analysis, and they rely on data engineers or others to handle coding tasks.

Our opinion is that the advantages of learning to code are so great that a Quant UXR without coding skills is at a long-term disadvantage. From the hiring perspective, we believe coding skills are highly advantageous even when a role may not require them. So our opinion is that a Quant UXR should, at a minimum, *try* to develop coding skills.

6.1.2 What Language?

If you program as a Quant UXR, with few exceptions, it should be in R or Python, possibly alongside other languages. R and Python are, by far, the most common languages used by Quant UXRs as well as colleagues in related roles such as data science. R and Python have the most support for handling data, performing statistical analysis, visualizing results, and testing and sharing code. You will be at a disadvantage if you don't work in at least one of them.

However, if you learn to program in another general-purpose language—such as C, C++, C#, Go, Java, JavaScript, Kotlin, Ruby, Scheme, or Swift, among others—then it will be easy for you to pick up R or Python. If any of those languages is preferable for your situation, we encourage you to develop skills there, and pick up R or Python when you're ready.

MatLab, SAS, Mathematica, Maple, Octave, Stata, and similar statistical and mathematical languages are special cases. Working with one of them, you might develop general programming skills, although you may need to devote special attention to the details we describe in the next section. Programming only in SPSS is mostly inadequate for Quant UXRs due to its limitations as a general-purpose language.

Finally, SQL is a special-purpose language for extracting and analyzing data from databases. This is a common task for Quant UXRs and many operations such as joining and filtering data are highly efficient in SQL. However, SQL is not a general-purpose language and thus is neither sufficient nor necessary for Quant UX work. We have included a section on SQL in this chapter to help you decide if you want to learn it.

6.2 Procedural Programming Basics

Let's assume that you've decided to program as a Quant UXR. What level of programming skill is required? Although there is no upper limit, there is a lower bound: you should be able to write *procedural* code for *general-purpose tasks* and especially *data handling*. We define *procedural programming* as applying algorithms to handle structured data and generate results, a combination that is reflected in the name of a famous computer science book, *Algorithms + Data Structures = Programming* [153]. In the following sections, we explain the essential elements of algorithms and data structures, specifically in regard to Quant UX.

6.2.1 Algorithms

For purposes of Quant UX programming, you should be fluent in three aspects of algorithms: converting a problem to a *logical sequence* of steps, perhaps using pseudocode (as previously noted); coding those steps in a language, using loops and other *control structures*; and generalizing the solution to a modular, reusable implementation as a *function* (also known as a procedure or subroutine).

6.2.1.1 Logical Steps

The conversion of a problem to a logical structure and pseudocode is a skill that comes almost entirely by practice. Introductory programming books, such as those mentioned in the "Learning More" section (Section 6.6), are helpful in presenting a sequence of very simple to progressively more complex examples. To develop the skill, you must personally *write* hundreds of such programs, either from texts or, better, in your own work. There is no way to learn to program without actually programming!

Pseudocode skills are closely related to *whiteboarding* skills, which refers to laying out a solution's overall logic and flow on a whiteboard (or paper, or some other medium) prior to coding. Whiteboarding is often requested of a candidate during a programming interview (see Chapter 12). We encourage you to write pseudocode whenever possible, to outline a solution before starting on the detailed code. We often do this directly in a programming editor, starting by writing a general outline of the approach in comments, and gradually converting small blocks into actual code.

6.2.1.2 Control Structures

To program effectively, you should be fluent in the logical mechanics of program flow, in particular if/then tests and for and while (or repeat until) loops. These are the absolute minimum foundations of fluency in programming, and any programming interview is overwhelmingly likely to focus on the implementation of logical steps using control structures.

A few common problems that arise in usage of control structures are

- Failing to consider zero-length input (for example, a for loop from 1 to length(X) will fail if X has zero length).

- Off-by-one errors, such as indexing a string from position [1] in a zero-indexed language, or incrementing a counter inside a loop before using the counter.

- Incorrect assumptions about data types. For example, consider the test `if(INPUT==0)`. What happens when `INPUT` is a character string?

During an interview, you should check the assumptions and state them carefully to demonstrate awareness. In general, candidates are not expected to explicitly write code to handle every possible edge case, but they are expected to consider a few possibilities and address one or two common problems.

6.2.1.3 Functions

Once you have line-by-line code that works, can you convert it to a generalized, self-contained function? A good function has several characteristics. First, it isolates a particular task, usually one that needs to be performed in several different places in code. By putting common code into a single function, it only needs to be updated in one place, and the program becomes more readable if you call a function instead of repeating very similar code in multiple places.

We develop such a reusable function for satisfaction charts in Section 8.4.2. You could refer to that to see how the function becomes advantageous in later sections. It is also an example of the level of programming complexity that you should expect for Quant UX analyses.

Second, a good function takes clear input parameters, returns a well-defined result that accomplishes its task, and has no other side effects. For example, a function should not change the value of anything other than its explicit *return value* (result). This is known as the *modularity* assumption of functions.

Third, a function should use clear names for itself and its input parameters. Consider a function whose purpose is to take a number such as 42 and convert it to a formatted currency string such as `"$42.00"`. A poor function declaration would be `convert(X)`. What is being converted? What is X? What does it return as output? A better definition is `convertToCurrency(xNum, to="USD")`. This communicates that the function performs a currency conversion, expects numeric input, and defaults to formatting the result as US dollars but also accepts some other options.

Fourth, a function should be paired with *test cases* that demonstrate its usefulness and confirm that it works as intended with both expected and unexpected input values. What would happen when we call `convertToCurrency(42, to="JPY")`, `convertToCurrency("Hello!")`, `convertToCurrency(1, 2, 3)`, `convertToCurrency()`, and `convertToCurrency(42, from="JPY", to="KRW")`? If you have clean data, a simple function that tests for only a few possibilities may be appropriate. But you should always consider the edge cases and handle the most likely.

Does this mean that you should set up explicit unit tests for your functions? It's a good idea, at least for important or widely used functions. *Unit tests* are checks that you design and include in your code to ensure that any function is giving the correct results for intended cases and edge cases that you want your code to handle. By making unit tests explicit, you can check that a function still works as intended after later code changes, such as optimization or bug fixes.

Finally, a good function has clear documentation, typically as code comments (and possibly written documents). Including comments is not necessary during interviews, but you'll thank yourself later if you add extensive comments during day-to-day coding.

Is that all? How about advanced fluency in algorithm design, efficiency, and solutions to typical problems such as traversing linked links or sorting? A few roles, especially those that overlap with data science or engineering, may require those skills, which veer into formal computer science. Our experience is that such skills are usually less important for Quant UXRs than general fluency in programming basics.

6.2.2 Data Structures

Computer science describes many kinds of data structures that are useful for various purposes, including vectors, arrays, linked lists, stacks, hash tables, trees, and directed graphs. Each structure is optimal for some purposes, suboptimal for others, and pairs naturally with certain algorithms (such as traversing a linked list or doing binary search of a tree).

Quant UXRs most often deal with two of these, *vectors* and *arrays*, plus a third hybrid type that is similar to an array, the *data frame*. Also important is the general concept underlying *hash tables*. We'll briefly describe each of these, and you can expect them to be important in a programming interview.

6.2.2.1 Vectors

A *vector* is a one-dimensional array comprising values of a single data type, such as (`2, 4, 10`), (`"a", "b", "c"`), and `501:505` (a sequence from 501–505). Vectors are important in Quant UX programming to keep track of how many times to do something, to select observations, and to count how many times something has been observed.

For example, using the first example (`2, 4, 10`), we might want to repeat the same calculation on each of the numbers 2, 4, and 10. From the third example, we might use the vector `501:505` to select observations 501 to 505 from a data sample.

The most obvious and easiest skill to develop with vectors is selecting data. For example, in R you might select observations from my.data using either directly specified or named vectors:

```
my.data <- data.frame(matrix(rnorm(10000), ncol=5))
my.data[501:505, ]
```

```
##                X1         X2         X3         X4         X5
## 501 -1.5595899 -2.1325391  0.7096897 -0.5663836 -0.13364897
## 502  1.2723982 -1.0330504 -1.1961304  1.8289368  1.80451982
## 503 -0.4406392  0.2545082  1.8022930 -1.3140974 -0.05186331
## 504 -0.3645085 -0.3537969  0.8219231 -0.7321998 -0.97498126
## 505  2.1178460 -0.7359523 -0.6294509  1.3619285  1.05892972
```

```
rows.select <- 501:505
my.data[rows.select, ]
```

```
##                X1         X2         X3         X4         X5
## 501 -1.5595899 -2.1325391  0.7096897 -0.5663836 -0.13364897
## 502  1.2723982 -1.0330504 -1.1961304  1.8289368  1.80451982
## 503 -0.4406392  0.2545082  1.8022930 -1.3140974 -0.05186331
## 504 -0.3645085 -0.3537969  0.8219231 -0.7321998 -0.97498126
## 505  2.1178460 -0.7359523 -0.6294509  1.3619285  1.05892972
```

A more difficult, yet fundamental skill is to *vectorize* operators, meaning that the code performs a task repeatedly by iterating over the values of a vector. This may be simple math, such as calculating several values at once in R:

```
c(2, 4, 10) ^ (1/2)    # square root of (2, 4, 10)
```

```
## [1] 1.414214 2.000000 3.162278
```

You may also vectorize a function call. The preceding example might be rewritten to apply a specific, user-defined function mathFunct to a vector object my.vector as follows:

```
my.vector <- c(2, 4, 10)
mathFunct  <- function(x) { sqrt(x) }
sapply(my.vector, mathFunct)
```

```
## [1] 1.414214 2.000000 3.162278
```

In this example, you can see the value of using a well-defined function, as previously described. The function mathFunct could be updated as needed—for example, to check for negative values—completely separately from the code that uses it later. Similarly, by using a vectorized operation (in this case, the R function sapply()), you are free to focus on the details of what you wish to accomplish, while the language handles the details such as the length of the object.

Do you need to master vectorization for coding interviews? For the simple cases— selecting data and doing vector math—yes. For the complex case of vectorizing arbitrary functions, no. In fact, you likely would be asked to write a for loop instead. Not only is that a basic skill to assess, it may also be a clearer and simpler solution. However, vectorization is an important skill to develop as a more advanced programmer.

6.2.2.2 Quick Check: What Do You Think?

If you're new to programming, what do you think so far? Does this chapter interest you? Have you accessed a copy of R to try some of the examples?

If it seems interesting but somewhat daunting, we again encourage you to try the examples hands on, and consider delving into a dedicated book on programming. On the other hand, if your response is one of indifference or even strong disinterest, then those are important signals, too—signals that programming might not be a good fit for you. People differ in skills and passions. Programming is not a good fit for everyone in practice, and simply feeling that you *should* learn it may not be enough to provide you with the persistence that is required to get through the learning curve.

6.2.2.3 Arrays and Data Frames

An *array* is a multidimensional object comprising values of one type, such as all numbers or characters. For Quant UXRs, the most common form of an array is a *matrix*, which some languages provide as a two-dimensional array with values of one data type. For example, in R, matrix(rep(NA, 10000), ncol=10) defines a two-dimensional array with a total of 10,000 elements, organized into 10 columns (and thus 1000 rows), each with a missing value (NA). Matrices are commonly used for mathematical operations, such as multiplying values to estimate linear models (commonly behind the scenes, not necessarily by users) and mapping conditional probabilities of events (an example, applied to web browsing activities, is given in [25], Section 14.4.1).

Data frames are similar to matrices and two-dimensional arrays insofar as they are two-dimensional objects organized into rows and columns. However, the columns are named and may have varied data types, such as observation identifiers (case IDs), categorical variables, numeric values, and character strings.

6.2.2.4 Hash Tables

A *hash table* has values similar to those in a vector, and adds an index field to select them. Whereas a vector or array is typically accessed through sequential, integer index values, such as [5] or [1:10], a hash table may be accessed with arbitrary values such as "Bill" or fbca6485ddce. Each entry in a hash table is a *key-value* pair, with a unique *key* that indexes the entry and a *value* that is obtained.

This kind of structure has different names in different programming languages. For example, in Python, a hash table is called a *dictionary*, and other languages use terms such as *map, hash, associative list*, and *keylist*. They all function similarly, and you should understand the concept in whatever language you prefer.

A common illustrative example of a hash table is a translation dictionary, where an original word (key) is paired with a translation (value). Another common example is a phone book, with names (keys) and phone numbers (values).

For interviewing, our recommendation is to become fluent in the *concept* and then find a particular implementation you might use during an interview.

6.3 SQL

SQL (Structured Query Language, often pronounced "sequel") is a special-purpose language for extracting and analyzing data from databases.

Web and mobile products almost always use a database as part of their back end (Section 2.1.2) and may also use one for product analytics. If you expect your work as a Quant UXR to involve analysis of product usage data, it's worthwhile for you to learn SQL.

To illustrate, imagine a social media product whose usage log data is stored in a database table called events, with fields including a user ID, a timestamp of when the event occurred, and an event type that corresponds to feature usage (logging in, writing a post, publishing a post, and so forth). With SQL, it is easy and fast to compute which features had the highest number of unique users in a given time period. For example:

```
SELECT event_type, COUNT(DISTINCT user_id) AS unique_users
FROM events
WHERE timestamp > '2022-01-01'
GROUP BY event_type
ORDER BY unique_users DESC
```

Let's take a look at that SQL code. First, the convention in SQL is to write keywords—built-in words that are part of the SQL language—in uppercase. In this query, the SELECT part identifies fields to show in the output. In that SELECT clause, the second element (COUNT) is an aggregator that counts the number of unique user IDs for each event type.

The WHERE clause filters the rows of the database to include only those in a given time range. GROUP BY tells the COUNT aggregator to separate its counts by event type. Finally, ORDER BY specifies that the output should be sorted according to the number of unique users and listed in descending order (DESC).

In a traditional programming language, those steps might take dozens of lines. SQL is very efficient and compact for basic data extraction and analysis tasks like these, and is also very fast for joining large data sets on the fields they have in common. Why not use SQL for everything? Because it is limited. Whereas SQL is specialized for working with data sets, general-purpose languages are far more flexible for a broad range of data manipulation, analysis, statistics, and visualization tasks. We strongly suggest that you learn a general-purpose programming language in addition to SQL. Our experience has been that general programmers can learn SQL quickly when it is needed.

6.4 Other Coding Topics

There are two other areas of programming with which you will want to have at least a passing familiarity: *reproducibility* and *performance*. Although Quant UXRs are not expected to be experts in either area, basic understanding is important and will help you both to grow as a programmer and to engage productively with colleagues.

6.4.1 Reproducibility of Code

Experienced programmers will already be familiar with the concepts of reproducibility, but if you are new to programming, you should pay some attention here. Note that this is separate from the reproducibility of analyses, although the two are related. Reproducible code is often a foundation of reproducible analysis.

Perhaps the most valuable aspect of programming is that your analyses, data handling, visualizations, and other tasks become *repeatable*. That poses questions about exactly how you manage that. What is your setup? How do you test your code? How do you share it with colleagues?

It is beyond our scope here to give detailed answers to those questions, yet we would highlight that they deserve careful attention. And, if you interview, you should expect to be asked about them. To get you started, we suggest the following directions:

1. What is your setup? Choose an IDE (integrated development environment, such as RStudio) or a notebook solution (such as Jupyter) and stick with it, at least until you start working at an organization that has standardized on something else.

2. How do you test your code? Learn about unit testing (see Section 6.2.1.3) and how to debug code in your setup environment. Also, consider when unit tests are helpful and not helpful (such as exploratory, one-time code).

3. Do you share code with colleagues? The canonical method here is to use a shared *Git* repository, which is commonly done through GitHub or a similar private system. Learning Git and GitHub is a good investment. Repository systems like GitHub track every change that is made to code and other files, and allow collaborators to work simultaneously, even offline. For individuals, such systems may be overkill, but as soon as you collaborate they become important or even indispensable. A good starting point is Tsitoara, *Beginning Git and GitHub* [142]. (By the way, we used a Git repository to share, back up, and collaborate on every word of this book.)

6.4.2 Performance and Scale

A good approach to programming is similar to pre-testing any research: start small, make sure it is working, and then scale up. For example, if you need to process 1000 large text files for sentiment analysis, a good approach is to start with a small fragment of one file. When your code works for that fragment, scale up to process a complete file. Then scale again to handle all 1000 files.

Before committing to that, you should ask whether it is actually necessary to scale up. If a data set is huge, you may not need to handle all of it. Often the best answer is to take a manageable random sample from a larger data set.

Suppose you want to answer a question such as "How many users use feature X?" One approach is to examine every user's records and count the occurrences of feature X. If you have millions or billions of users, this may be prohibitively slow! Another approach is to sample a small proportion, perhaps in the range of only 10,000 or 100,000 users. Find the proportion of them using feature X, and multiply by the total number of users. Similarly, if you are fitting statistical models, there is rarely a need to model more than 100,000 users (sets of observations) because you'll usually be interested in effects that are large enough to be detected at that size (see Section 5.3.3).

Some problems, however, do require processing massive amounts of data. A common case is pre-processing data so that it is ready for further analysis. For instance, suppose that usage of feature X is not recorded on its own but must be inferred from some sequence of events or other observations. You may need to process all of the data first to identify usage of feature X, and then consider sampling it.

When massive scale is required, as opposed to merely sampling, you will want to consider three things:

1. Is the code reasonably well-optimized? You should invest time to do *code profiling* and identify bottlenecks in the code. A code profile shows how much time and memory are taken by each line, section, or function in code. You can then optimize the parts that are slow or that use excessive system resources.

2. Have you chosen an algorithm that scales well, preferably— although this is not always possible—at less than a linear rate? (Meaning, if you apply it to 100× as much data, it will take *less than* 100× as much time.) Algorithmic *complexity* is out of scope here and we generally refer Quant UXRs to talk with computer scientists when such issues arise. If you are interested in this, see the previous discussion about vectorization (Section 6.2.2.1) and the "Learning More" section (Section 6.6).

3. Can the problem be solved by a simple divide-and-conquer approach? A simple way to solve many problems is to distribute them across a *cluster* of machines or processing cores. Cloud

services such as Amazon Web Services, Microsoft Azure, and Google Cloud make it possible to do this. A key concept is the *MapReduce* approach. First, the problem is solved once, and that solution is *mapped* out to occur many times in parallel on many processing units. Then the results are gathered from those parallel sources and *reduced* to give the result you need. Consider the previous example of 1000 sentiment text files. You could map out processing each of the 1000 sentiment text files to a computing cluster, gather processed sentiment scores from each of them, and reduce those scores to a summary or other statistical analysis as needed.

As with the other advanced programming topics discussed in this chapter, it is not essential for Quant UXRs to be experts in code performance and scale. Rather, they should understand the concepts and be able to consider possible solutions to handle massive data when they may be confronted by it.

6.5 Key Points

Although Quant UXRs differ in their opinion of whether programming skill is required for Quant UX, they agree that it is at least desirable and advantageous. We recommend that Quant UXRs be fluent in the following aspects of programming:

- A language such as R or Python that is commonly used for data analysis (Section 6.1.2)

- An interactive environment for that language

- Common packages or libraries for handling data

- Writing reusable functions and testing their input and output (Section 6.2.1.3)

- Handling common data types, especially data frames (Sections 6.2.2, 6.2.2.3)

- Replication of code results (Section 6.4.1)

- Basic concepts of how to scale code to handle very large data (Section 6.4.2)

By contrast, some of the advanced skills that are not required for most Quant UXRs are

- Interaction between languages, such as using R together with C++

- Writing packages and libraries for others to use

- The full range of computer science algorithms

- Detailed optimization of code

- Fluency with cloud computing systems

Those advanced skills are certainly valuable if you are interested in them, although you may consider instead whether to develop additional skills in human-centered research (see Chapter 4, "UX Research") or statistics (see Chapter 5, "Statistics").

6.6 Learning More

If you are a *beginning* programmer interested in analytics, we recommend that you start with either R or Python, the two most common languages used by Quant UXRs, data analysts, and many others. The choice between them depends largely on whether you are more interested in statistics (R) or general data handling and engineering (Python). Those are covered by the companion R [25] and Python [127] texts.

There is an alternative path that a beginner might consider: learn to program in a general-purpose language that has little to do with analytics. The advantage of this approach is that you will learn general programming skills that apply to any context, and will not have to consider details of statistical models or analyses while doing that. The disadvantage is that it may be less motivating and less applicable to immediate analytic work. A classic text for this approach—and a wonderful read for any programmer—is *The C Programming Language* [68].

For *intermediate* programmers, you will want to develop your coding skills further. For Python, a great second book that helps you to think like a Python programmer and solidify your skills is *Beyond the Basic Stuff with Python* [138]. In R, a similar book that teaches many best practices, explains how R works, and teaches basic computer science along the way is *The Art of R Programming* [90].

To be a successful Quant UXR, it is generally unnecessary to go beyond the intermediate level of programming skill. However, if you are highly motivated and want to keep going, we have two additional recommendations. *Cracking the Coding Interview*

[93] provides a crash course across a vast range of programming questions (definitely not for beginners!) If you are interested to learn more about computer science and formalize your understanding of algorithms, Sedgewick and Wayne's *Algorithms* is an outstanding introduction [128]. Unfortunately for Quant UXRs, it uses Java as its base language—but don't let that stop you. If you program in Python, R, or any other language, you will find Java to be readable.

6.7 Exercises

Although most introductory programming books contain many problems to solve, they are generally remote from the activities of Quant UXRs. We present the following two exercises that are similar to problems Quant UXRs might solve with code, and demonstrate the approximate level of minimum programming skill that we recommend for Quant UXRs.

1. Translate License Plates

   - Write a function that will take a typical license plate string and convert it to the NATO phonetic alphabet (`https://en.wikipedia.org/wiki/NATO_phonetic_alphabet`). For example, the license plate input "`QUANT 1`" would give the result "`Quebec Uniform Alpha November Tango One`".

   - Now write a function that will generate 1000 random license plate strings of four to eight characters. Translate all of them to their NATO phonetic equivalents.

     Edge Cases. What does your function do for the following input values? "", "`QU@NT`", and `NULL` (a null value as given in R; varies by language).

2. Interleaving Two Files

   - **Part 1**. Write a self-contained, named function (Section 6.2.1.3) that takes two file names. Assume that the files are standard text files with the lines in sorted order, according to standard English alphabetical sorting. Combine the files, one line at a time, into a single output file that remains in sorted order.

- **Part 2**. Ignore any line that begins with the "#" character, and simply copy it to the output file wherever it appears. The function must process only one or two lines at a time from either file; assume they are too large to fit entirely into memory.

- **Part 3**. Now write a function that accepts *any* number of file names—perhaps 5, 10, or 1000 files—and combines output in the same way, to list each line in sorted order regardless of which file it came from.

- *Questions*. What happens if a file doesn't exist? What if the files have different text encoding? Suppose you need to combine 100,000 files that each have 1 billion lines, and obtain a result within two hours. How would you approach that?

PART III

Tools and Techniques

Introduction to Part III

In this part, Chapters 7 through 10, we take a less general approach than in other parts of this book. We do not attempt to describe the most common or most important methods used by Quant UXRs, but rather ones that are most *distinctive* and additive to information you can find elsewhere. Each chapter discusses concepts, methods, or code that is uniquely presented here.

Among these four chapters, the first two should be of interest and value to nearly any UX researcher who is interested in quantitative methods. Chapter 7, "Metrics of User Experience," discusses the straightforward and yet powerful *HEART* framework to define metrics for assessing user experience. Chapter 8, "Customer Satisfaction Surveys," examines one of the most common sources of data for UX researchers, customer satisfaction surveys. It discusses best practices, challenges, and potential pitfalls from a perspective gained through hundreds of survey projects.

The second two chapters are more specialized and include intensive R code. In Chapter 9, "Log Sequence Visualization," we examine behavioral logs that might be collected in a product or website, and demonstrate how to aggregate sequences of user actions into interactive *sunburst* visualizations. Chapter 10, "MaxDiff: Prioritizing Features and User Needs," discusses the *MaxDiff* survey method that is rapidly gaining popularity among UX researchers. MaxDiff allows Quant UXRs to apply their skills in the earliest stages of product planning and definition.

CHAPTER 7

Metrics of User Experience

User experience metrics are an important topic because they relate to the most common reason UX teams want to include a Quant UX researcher: to measure the impact of UX work on a project. Large-scale A/B testing (Section 5.3.4) makes it possible to compare designs to each other, and for UX teams it is essential that these comparisons are done in a way that reflects the quality of user experience.

By default, designs will end up being evaluated against whichever business metrics are in place. Such metrics are often related to user experience, given that a product with poor UX would be unlikely to succeed as a business. But this relationship is usually indirect, and UX improvements may not lead to an obvious or immediate change in key business metrics. In some situations, a better design for users may end up performing worse on these default metrics, leading to frustration and second-guessing. For example, an improvement to the search function on a website may lead to a decrease in page views per user, because users can find what they are looking for with fewer clicks. With intention and focus, Quant UXRs can do better than this, finding proxies for user experience in survey or usage data.

In this chapter, we talk about ways in which Quant UXRs can use their unique skill set to help teams: to figure out whether designs are working for users, and to get on the same page about the purpose of their project. This includes two useful methods: the HEART framework (to choose metrics that reflect the quality of user experience) and the Goals-Signals-Metrics process (to develop metrics that reflect the goals of your product or project).

These methods were developed by the earliest Quant UXRs at Google, led by Kerry, and have been in use there for more than 10 years. They have also been adopted by many other organizations. We close the chapter with lessons learned from using these methods in practice.

© Chris Chapman and Kerry Rodden 2023
C. Chapman and K. Rodden, *Quantitative User Experience Research*,
https://doi.org/10.1007/978-1-4842-9268-6_7

7.1 The HEART Framework

The early Quant UXRs at Google were often asked to help product teams define metrics of user experience, and over time, they noticed that their suggestions tended to fall into five categories, which they formed into an acronym for ease of recall: Happiness, Engagement, Adoption, Retention, and Task success (HEART).

It is helpful to use these categories to break down the broad area of "user experience" into more specific concepts that reflect the outcomes the team is working toward—for example, what user benefit does the team expect to see from a redesign of a given feature, and how would that be reflected in user attitudes or behavior? We say more about the process for this in Section 7.2, but first we'll describe the HEART categories and how those relate to user-centric project outcomes.

7.1.1 Happiness

Happiness is a blanket term for measures of user attitudes, often collected via survey. This could include satisfaction, perceived ease of use, and targeted questions about specific aspects of the user experience that would be hard to gather from behavioral data.

The term "Happiness" came from "Happiness Tracking Surveys (HaTS)" [95], a Google initiative to deploy consistent user experience surveys across products and provide teams with a resource to gather ongoing signals of user satisfaction. The metrics, of course, are not representative of a user's happiness in the broader sense. If users are satisfied with your product, that doesn't necessarily mean that the product is improving their lives (see Section 4.5).

7.1.2 Engagement

Engagement is the degree of a user's involvement with a product, typically measured via behavioral proxies such as frequency, intensity, or depth of interaction. For example, within a given time period, a team might choose to measure the number of visits made, or the amount of usage of a key product feature (such as number of photos uploaded to a photo-sharing product).

Engagement is useful to measure because some percentage of users are just casually checking out a product, while others are actively involved with it, and that distinction is an important one for long-term product success.

A product generally has users at different tiers of engagement, from very casual to very deep, and so an increase in average engagement across all users may require further analysis in order to interpret the impact. For example, did your change encourage the most involved users to start using the product even more, or did it encourage casual users to move to a higher level of involvement? (The latter may be a more useful long-term outcome, depending on your product's goals.)

7.1.3 Adoption

Adoption refers to counting new users of a product or feature—making an explicit distinction between new and existing users, not just combining them in a single number. It's easy to count the raw number of unique users of a product during some time period, and product dashboards commonly contain metrics such as daily active users (DAU) or monthly active users (MAU).

Separating out new users from existing users enables you to understand how quickly your user base is growing, which is especially important for new products or features. Typically, you will use account signup or app install as a method of identifying new users, but it may also be useful to define some further action that you will count as more active product adoption; for example, in Google Docs you might consider that adoption has happened only after the user has created their first document.

7.1.4 Retention

Retention is the rate at which users are returning to the product, and can be thought of as a longer-term version of engagement. A team might, for example, choose to measure the percentage of users from a given week who used the product again in the following week. If users are finding a product valuable, they should continue to return to it over a long period of time. Some teams focus more specifically on failure to retain, which is known as "churn."

A common approach to understanding retention is known as *cohort analysis*: take a group of users who all signed up in the same time period (a cohort) and follow their usage across subsequent time periods to see whether they continue to return to the product.

7.1.5 Task Success

Task success relates to traditional behavioral metrics of user experience, such as those that would be measured in a usability lab (see Section 2.2). This includes efficiency (e.g., time to complete a task), effectiveness (e.g., percent of users who complete the task), and error rate.

This category often leads to metrics that are the most specific and sensitive with regard to UX changes, but it can be challenging to collect the necessary data on a large scale. This is because you need to able to track events that are reliable indicators of the task being started, completed, or abandoned. It is easiest to do this in areas of your product where users do well-defined tasks, such as search or an upload flow. For example, if a user clicks an Upload button, this event marks the beginning of a task whose progress can be tracked to completion (or abandonment) via logs of the subsequent known steps in the flow. This enables measurement of the percentage of uploads that were successful, and also time-to-completion if the events are timestamped. It is also common to visualize the percentage of users who abandon the task at each step of the process, which is known as *funnel analysis*.

Some types of tasks are more complex and do not map easily to events in log data. An option in this case is to use assigned tasks in a large-scale version of a usability study, sometimes known as *benchmarking*. Like surveys, this approach generally requires use of a paid third-party platform.

7.2 The Goals-Signals-Metrics Process

It's tempting to start thinking about metrics by simply brainstorming a long list, inspired by the HEART categories, but this can quickly get unwieldy and hard to prioritize. Ideally, you want a small set of key metrics that everyone on the team cares about. To figure out what those are, you need to start at a higher level: identify your goals so you can choose metrics that help you measure progress toward those goals.

7.2.1 Goals

It can be surprisingly hard to articulate the goals of a project, even for the people who work on that project. At Google, many conversations in UX metrics office hours would go like this:

Designer: "We're building a dashboard! Can you tell us what our metrics should be?"

Quant UXR: "Well, what are the goals of your project?"

Designer: "Um…let me get back to you on that."

It's common at first for team members to try to state goals in terms of existing business metrics, like "increase traffic to our product." Through discussion, you can start encouraging them to be more precise. For example, is the team focused on ways of making the product more engaging for existing users, or on making it easier for new users to get started?

This part of the process is where it is helpful to bring in the HEART framework, as described in the previous section. HEART helps influence a product team to stay focused on goals that relate to improving user experience.

In many cases, different members of a team have different ideas about the user experience goals of a project, and they may not realize it. For example, imagine a team that is redesigning a navigation bar. The PM may think that the primary purpose of the redesign is to make room to promote useful new features; the engineers may see it as an opportunity to make the animations smoother; the UX designer may be focused on reorganizing the categories so that they map more closely to the expectations of new users; and the UX researcher may be concerned with making sure that existing users are not confused by the changes. An explicit discussion of goals, in the concrete context of deciding how to measure the success of the project, provides an opportunity to expose these different points of view and work to build some consensus around priorities. So it's important that defining goals is part of a team effort, not something that a UX researcher (or any team member) does alone.

Let's look at potential goals for a video-sharing app. A high-level goal might be, "Users watch videos and enjoy them." This actually breaks into two goals: one for *engagement* (watching videos) and one for *happiness* (enjoying them). The HEART categories can also be applied at the level of a specific feature; the product team might be interested in measuring *adoption* of key features like sharing videos, as a further indicator of enjoyment. Finally, the search feature would have a *task success* goal: users should be able to quickly and easily find the videos that are most relevant to their query.

It is important to start with goals at this kind of high level—as opposed to beginning with something like "increase viewing time by 10%"—because it provides the opportunity to explicitly bring in user experience considerations, and opens the team to multiple possible ways of measuring progress toward a goal.

7.2.2 Signals

After you agree on product goals, you can map those goals to possible lower-level signals. How might success or failure in the goals manifest itself in user behavior or attitudes? For example, returning to the video-sharing app from the previous section, an engagement signal might be the number of videos users watch, or the amount of time they spend watching those videos. A task success signal for the search feature could be whether a user clicks any of the results.

There are usually a large number of potentially useful signals for a particular goal. Once you have generated promising candidates, you will need to pause and do some analysis of preliminary data, followed by discussion with stakeholders, to help you decide which ones appear to be most promising and useful—and to convince yourself that they correspond to an improved user experience. For example, you could determine whether a particular short-term engagement signal predicts long-term retention or measured satisfaction.

When choosing among possible signals:

- Consider how easy or difficult it is to track each one. Is your product instrumented to log the relevant actions, or could it be? If your candidate signal relates to user attitudes, could you deploy an in-product survey on a regular basis to measure changes over time?

- Consider how sensitive the signal will be to changes in your design. If you're already collecting potentially useful signals, you can analyze the data you have and try to understand which signals seem to be the best predictors of the associated goal. If your team is redesigning a specific feature, you may want to focus on signals relating to that feature in particular, because it is less likely that your redesign will lead to changes in overall product usage or satisfaction, for example.

Note at this point that a *signal* is not yet a measurable *metric*. A signal translates a goal (or more commonly, *part* of a goal) into a general area of measurement, but it does not yet detail a precise metric to implement on a dashboard. A given goal is likely to have more than one possible signal; and a signal will have more than one possible metric, as you'll see next.

7.2.3 Metrics

Once you've chosen signals, you can refine those further, into metrics you'll track over time on a dashboard, or use for comparison in an A/B test. Continuing with the example of the video-sharing app, you might implement the general signal "how long users spend watching videos" as a more specific metric of "the average number of minutes spent watching videos per user per day" (although with ethical considerations, discussed in Section 7.5.2.3). You could implement the signal of "whether the user clicks on any of the search results" as the more specific metric, "the percentage of first-page queries where the user either takes no further action, goes to the next page of results, or changes their query to something else" (note that this would be a metric of search failure, rather than success).

Let's look again at the distinction between signals and metrics. You can think of the signal as the data you need to collect, and the metric as how you will analyze that data to turn it into a number or chart on your dashboard. There will be many possible metrics you could create from a given signal—you'll need to do some analysis of the data you've already collected to decide what's most appropriate, such as (in the search example) what exactly should count as "whether they click any of the results."

You will need to normalize raw totals or counts to make them more meaningful, for example, by using averages or percentages. You will also need to aggregate over a given time period (like a day, or a week) and track how the metric changes from one period to the next. Analyzing by month is tricky because of the different number of days in each month; it's common to use 30 days as an alternative, but it's better to use a 28-day period to avoid inconsistencies caused by where weekends fall.

The landscape of metrics tools such as dashboard engines is large and constantly changing, so we will not make recommendations here. It should be clear by now, however, that the built-in metrics in off-the-shelf analytics tools are unlikely to meet all of your needs, because the most useful metrics will be specific to your product or project.

At this point, when you are deep in the details of metric implementation, it's tempting to go beyond your original list and start adding "interesting stats" that don't necessarily relate directly to the original goals you defined. Before you do that, remember that every new chart adds clutter and cognitive overhead for whoever is looking at the dashboard, creates maintenance overhead for your colleagues or your future self, and risks stakeholders getting distracted or selectively choosing misleading data points. Consider starting with a one-off analysis of your new idea, and get feedback from the team about its value and whether it makes sense to track it over time.

7.3 Applying the Methods Together

So, now you know about a couple of conceptual frameworks—HEART and Goals-Signals-Metrics—but the important part is applying them in practice. In this section, we examine practical considerations related to applying these methods in real projects. They work best as the basis for facilitating a focused discussion with your team.

The Goals-Signals-Metrics process can be applied to any situation where you need to develop metrics, and is not specific to user experience metrics. The UX aspect is where the HEART categories come in; when discussing the goals of your project, you can run through the HEART categories as a shortcut to help you ensure that you are including goals related to user experience. The HEART acronym was developed to be memorable, so it is easy to recall the categories in a meeting and write them on a whiteboard. However, HEART is only a useful starting point. You may have goals that don't fit into the existing categories, and that is fine. If you can't fit a goal into one of the categories, that doesn't mean it isn't valid; don't try to force it. The important thing about HEART is to consider all five categories, and then keep or add the ones that are important for your product and goals.

For any project there are a very large number of possible metrics that you could track, and it would take too much effort for your team to implement them as well as too much attention for anyone to review or understand all of the resulting charts. The Goals-Signals-Metrics process should lead to a natural prioritization of the various metrics. It's most important to track the metrics related to your top goals.

Also, not all of the categories will apply to every project. For example, if you work on an enterprise product that people must use as part of their daily work, it may not make sense to consider product-level engagement, adoption, and retention. The users themselves may have little choice as to whether to adopt and use the product. Instead, a team may choose to focus more on happiness or task success. They could also consider feature-level engagement, adoption, and retention, which may indicate the utility of features for specific subsets of users. For example, you might want to measure the percentage of users within an organization who adopt a feature within the first week after it is released, or the percentage of active users who use a feature in a given week.

7.4 Example: Redesigning Labels in Gmail

Let's look at a real example of applying HEART and Goals-Signals-Metrics, from one of Kerry's early Quant UXR projects at Google [119].

Gmail, Google's web-based email product, launched with the feature of *labels* for organizing mail. Compared to folders, which were the traditional method, labels are more flexible: an email conversation can be given more than one label; for example, a flight booking confirmation could be labeled both "Travel" and "Receipts." However, the feature was difficult to discover, leading to the mistaken impression among many users that Gmail did not offer features for organizing mail; the most common feature request was that Gmail should add folders.

To address this, the Gmail team launched a redesign of the labeling features in 2009, to make them more visible:

- New "Move to" and "Labels" drop-downs were added to the toolbar, pulling the functionality out from the "More actions" drop-down (see Figure 7-1). "Move to" was designed to offer folder-like behavior of simultaneously applying a label to a conversation and removing it from the inbox. The idea was that this should feel familiar to a user who expects folders, without them having to worry about any new terminology. The "Labels" drop-down provided a more visible way to access the existing functionality of simply applying a label to a conversation.

- Labels were given a more prominent position in the left navigation bar, displayed next to the system labels like "Drafts" and "Sent Mail."

These two aspects launched separately, several months apart, but for simplicity we will consider them here as a single redesign. They were not subject to A/B testing because the changes were big enough to require user education and marketing, so the analysis was based on comparing label usage before and after the redesign.

This project was considered a priority by the UX team in particular, because it was driven by user needs and feedback. Its design process was thoughtful and thorough, involving many iterations, prototypes, and user studies, which took a lot of UX team effort. So when it launched, it was especially important to be able to show that the effort had been worthwhile, in terms of actually helping people discover labels to organize their mail.

To illustrate why HEART and Goals-Signals-Metrics are helpful, let's consider for a moment what the analysis might have looked like without them. The simplest possible chart of label usage, given how most logging systems are set up, would be a time series of the overall number of labeling actions (either adding or removing a label from a conversation), as shown in Figure 7-2.

We might see a bump at the time of the launch, and the number of labeling actions going up over time. Does this indicate that the redesign was a success?

Figure 7-1. *In a redesign of its labeling features, Gmail introduced new "Move to" and "Labels" drop-downs*

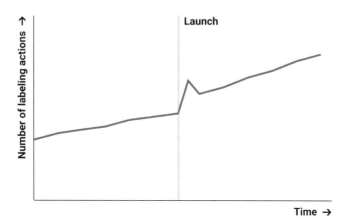

Figure 7-2. *A not particularly useful way to look at the impact of the redesign: number of labeling actions over time*

Not really. This chart doesn't tell us much that is specific to the intention behind the changes that were made. In particular, we cannot tell whether the increase over time is because

- A higher percentage of Gmail users are using labels

- Gmail users who are already using labels are using them more often

- Gmail is adding more users over time

So let's now recreate how the team applied Goals-Signals-Metrics, first using HEART to clarify the goal of the redesign: for a user who wants to file their mail, the goal was to help them discover that they could do it with labels. So Adoption was the most relevant HEART category: the team wanted more Gmail users to discover and use labels to file their mail.

Following the rest of the Goals-Signals-Metrics process:

- *Goal*: More Gmail users discover and use labels to file their mail

- *Signal*: Number of Gmail users who use labels (where "use" is defined as adding or removing a label from a conversation)

- *Metric*: Percentage of 7-day-active Gmail users who added or removed at least one label during those 7 days

Note that there was no target for this metric, because the team did not know what percentage of users wanted to file their mail, but given that there was a problem with discovering the existing labeling features, the team were looking for some increase in the percentage of Gmail users who used labels. A user survey could have helped establish a more precise target, if needed.

With this new choice of metric, the analysis looked more like Figure 7-3. The combination of both launches led to an 80% increase in the percentage of Gmail users who used labels, relative to before the redesign.

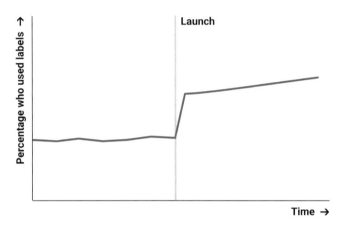

Figure 7-3. *A more useful chart, after defining a metric that is specific to the goal of the redesign: the percentage of 7-day-active Gmail users who added or removed at least one label during those 7 days*

It's often important to also consider impact on a particular subset of users, and in this case the team looked specifically at new Gmail users, and found that after both launches they were about twice as likely to apply a label in their first two weeks after signup, compared to the old designs.

Although Adoption was the primary goal, the team also considered metrics from other categories, as a safeguard:

- *Engagement*: The number of labeling actions per user remained the same over time. This was the expected result, given that this metric should be largely driven by the underlying number of email conversations that users need to file, which should be unaffected by the redesign.

- *Task success*: At the time of the launch, there was an increased incidence of undo-like actions associated with labeling (e.g., removing a label immediately after it was added), perhaps indicating that users were making more errors. This quickly returned to its previous level, so there was no cause for concern.

7.5 Lessons Learned From Experience

The concepts of HEART and Goals-Signals-Metrics were first developed more than 10 years ago, and in the meantime, many teams inside and outside of Google have found them helpful and have applied them to a large array of products and services. From these experiences, we've learned a lot about making UX metrics work in practice. In this section we address potential pitfalls and broader organizational issues that can impede successful application of these concepts, along with our suggestions on how to overcome them.

7.5.1 Individual Pitfalls

We start with some issues that you might encounter as an individual researcher while trying to apply these concepts.

7.5.1.1 Not Enough Team Involvement

It's tempting to think that you can go through the Goals-Signals-Metrics exercise on your own and send around a summary to your team for feedback. This can feel easier and faster than trying to find time on team members' calendars to do the exercise with them. However, this will compromise the impact of any resulting assessments, because your team members will feel less ownership over them. It will also lower the quality of your metrics ideas, because you will not benefit from the engaged input of the team.

7.5.1.2 Starting Too Big

The HEART framework is intended to be used with individual teams who are thinking carefully about the details. It probably will not work out well if you try to create a "HEART dashboard" for an entire organization. Start small, with one project and a team that is receptive. Pay attention to the details and learn from the experience. Then you will have an internal case study that other teams can learn from.

7.5.1.3 Underestimating the Next Steps

HEART helps the most with the first part of the work: brainstorming, considering, refining, and agreeing on high-level goals with your team, and then working through detailed ideas about signals and metrics. But after that, you are not done! You still have to go through the process of analyzing the data, iterating on your signals and metrics (while keeping your team in the loop), and figuring out how you will implement your chosen metrics. All of that is a process that takes time.

7.5.1.4 Too Many Metrics

The HEART prompts can be very useful for brainstorming about goals, signals, and metrics, but there is a risk that you and your team will get so excited about the ideas that you will try to implement all of them at once. It is crucial to be clear about which metrics are your highest priority and to make a distinction between these and any other metrics, for example, when creating a project dashboard or conducting A/B tests.

Avoiding that effort means that you will end up with a giant dashboard that is overwhelming to look at, especially for stakeholders who are only interested in knowing the key outcomes. In the context of A/B testing, using too many metrics also increases the risk of a false positive result, because the more statistical tests you do, the more likely it is that one of them will be significant just by chance. You will, of course, need to make use of other metrics to help you dig deeper into particular questions—for example, to debug what went wrong with an experiment—but that does not mean that you need to implement all of them upfront or in the same place.

We will reiterate here that just because HEART has five categories, you don't need to use all of them. Pick only the categories that are most relevant to your project.

7.5.2 Organizational Issues

These are broader issues relating to the team or organization, which often appear when developing and implementing UX metrics.

7.5.2.1 Unwillingness to be Evaluated

Any time you introduce new metrics, especially if they are specifically targeted in the way we advocate here, you increase the chance of exposing that a project was unsuccessful. Team members and leaders need to feel safe accepting that risk, which may be difficult if your organization does not have a culture of openly discussing and learning from failures (e.g., blameless post-mortems). The issue may not come up during the definition phase, but afterward, when it's clear that the metrics are not "up and to the right," be aware that you may encounter defensiveness and questioning of the value of your work.

This is another reason to ensure that you involve the rest of your team (including key stakeholders) in the process of defining and implementing metrics: if they trusted the process and felt some ownership over it, they are more likely to trust and accept the outcomes, even if negative.

7.5.2.2 Optimizing for a Single Metric

In the previous section we noted the pitfall of "too many metrics," but some organizations go too far in the other direction and choose only one metric.

Leaders often have a desire to rally their organization behind a single metric, to provide clarity and focus. However, teams may then make the implicit (and incorrect) assumption that positive changes in this primary metric mean that everything is going well with the product. As we discussed in Section 5.4.3, teams need to use multiple metrics. Metrics are often not correlated with each other, and the other key metrics are essential checks and balances—a team must continue to monitor them and understand the trade-offs that exist between them.

This is especially true when the primary metric becomes an optimization target (for example, using machine learning), because this renders the primary metric useless as a means of evaluation. This particular pitfall is well-known enough that it has a name, *Goodhart's Law*, phrased by Marilyn Strathern [137] as, "When a measure becomes a target, it ceases to be a good measure."

7.5.2.3 Failure to Consider Ethical Consequences

Although the HEART categories are intended to help measure the quality of user experience, the metrics derived from them are only proxies for users' actual experiences. Ethical considerations require an organizational willingness to define higher-level values and appropriate boundaries, focusing on longer-term outcomes. In Section 7.2, we gave a running example of using time spent watching videos to measure engagement with a video-sharing app. This does not measure true engagement, however, and has many limitations, including a short-term focus. At how many minutes per day does engagement cross over into compulsive use that has negative consequences on a user's life? If a user believes that they are spending too much time in a product, might they stop using it altogether? Qualitative research can provide insight into questions like these and help the organization get a better understanding of what actually constitutes a good user experience, and iterate on key metrics as needed.

As we have emphasized throughout this chapter, a thoughtful approach to metric selection requires consideration of product-specific nuances, not simply reusing the same set of metrics across products. For example, in Gmail it does not make sense to think of engagement in terms of either time spent or absolute frequency of checking email. Few users would prefer to spend more time on email. Good product design would make email more efficient, not less. Instead it may be more meaningful to think of engagement in terms of how many days in a week the user checks their account (which has a low ceiling of 7) or whether users choose to send mail from the account in a given week. It may be even more meaningful to focus on other metric categories such as retention or task success.

We say more about research ethics in Section 4.5.

7.6 Key Points

In this chapter, we introduced the Goals–Signals–Metrics process and the HEART framework for UX metrics. These are useful tools to help teams collaboratively define large-scale metrics of user experience.

- A common responsibility for a quantitative UX researcher is measuring the impact of UX work on a project, and this requires defining specific metrics of user experience for your project (Section 7.2).

- It is crucial to work with your team to agree on the primary goals of your project, with regard to user experience. Refine those goals to signals, and then refine the signals to metrics (Sections 7.2.2 and 7.2.3).

- The categories in HEART (Happiness, Engagement, Adoption, Retention, and Task success) are a useful reminder of major types of goals that relate to a positive user experience (Section 7.1). The HEART acronym is helpful to recall the categories while leading an informal discussion.

- This work is complicated and nuanced in practice, and doing it effectively is just as much about organizational dynamics as it is about quantitative analysis. There are many pitfalls along the way, and we recommend to start small and grow the scope over time as you gain familiarity with the process and have success stories to inspire colleagues (Section 7.5).

7.7 Learning More

The HEART framework and Goals-Signals-Metrics first appeared in a paper by Kerry and colleagues, "Measuring the User Experience on a Large Scale: User-Centered Metrics for Web Applications" [118] at the ACM CHI 2010 Conference.

The Gmail labels example is written up in more detail in a case study paper from the same conference [119].

Some HEART categories have much more published literature than others. In particular:

- The *Happiness* category corresponds to large-scale measures of user attitudes, most commonly collected via surveys such as Google's Happiness Tracking Surveys [95]. Many other survey instruments have been developed over the years to measure user experience, summarized in Chapter 8 of *Quantifying the User Experience* [124]. Customer satisfaction surveys are discussed in the next chapter of this book (Chapter 8, "Customer Satisfaction Surveys").

- The *Task success* category overlaps traditional smaller-scale behavioral measures of user experience, such as efficiency and effectiveness metrics used in lab-based usability studies. A good general text for these is *Measuring the User Experience* by Tullis and Albert [144].

Engagement, Adoption, and *Retention* are newer classes of metrics that depend on the availability of large-scale usage data, so they have fewer reference sources to date. Similar concepts are discussed in texts aimed at startup founders, such as *Lean Analytics* [36]. Chapter 9, "Log Sequence Visualization," in this book discusses log analysis, which—depending on the product and data sources—is often used for assessment of Engagement, Adoption, and Retention. This would be a great opportunity for you, the reader, to share case studies with the Quant UX community [30].

For more depth on the specifics of using metrics in practice, especially in the context of A/B tests, we recommend Kohavi, Tang, and Xu's *Trustworthy Online Controlled Experiments* [72]. Chapter 6 of that book makes a useful distinction between *goal metrics* (used to track progress toward an organization's long-term "North Star" goals), *driver metrics* (shorter-term proxies for the goal metrics, such as an engagement metric that is a proxy for long-term retention), and *guardrail metrics* (to quickly identify when something has gone wrong with an experiment).

7.8 Exercises

The first exercise is an introduction to applying the tools in this chapter to defining user experience metrics for a product. The other two are intended to get you thinking beyond the HEART framework to other categories of goals that a product team might have.

1. Think about a product that you use often—does it have a feature that you'd like to redesign? If your changes could be implemented, how would you know whether they were successful, in terms of improving the user experience?

2. The HEART framework was mostly created with consumer products in mind, and we mentioned some adaptations for enterprise applications. How might you adapt it for online government services, like applying for benefits or renewing a driver's license?

3. The HEART framework focuses on an individual's experience of using a product. What additional considerations are there for a product or feature whose use cases are primarily social? For example, how would you measure the impact of a change to the way that comments are displayed in a photo-sharing app?

## CHAPTER 8

Customer Satisfaction Surveys

Customer satisfaction (CSat) surveys are some of the most common projects for Quant UXRs. In many cases, CSat surveys are written, fielded, and analyzed by Quant UXRs. In some organizations, CSat surveys are conducted by survey scientists or marketing researchers, but Quant UXRs are asked to review or collaborate on the projects and need to understand the underlying principles.

In this chapter, we provide guidance for writing, fielding, and analyzing customer satisfaction surveys. Along the way, we review common problems and misunderstandings. In Section 8.2.7.1 we summarize the key things to report. This chapter also provides hands-on code and data. We present R code to perform core analyses and use it to illustrate a frequent problem in CSat projects.

This chapter is sufficient to get started with CSat projects, but there are several topics you will want to consider for additional learning, including statistical concerns and text analytics. We point to resources for those in the "Learning More" section (Section 8.6).

Note that we often use the term *customer* in this chapter instead of *user* because that is the term most commonly used in these programs. These surveys and methods might be used with paying customers, individual users, or any other appropriate sample, depending on the research goal.

8.1 Goals of a Customer Satisfaction Program

The most important goal of a customer satisfaction program is this: *listen to and learn from your customers.*

Does that sound obvious? In practice, we often find that it is *not* obvious to stakeholders, who may believe that the goal is to detect change between releases, to compare to competitive benchmarks, or to identify customer groups who have specific problems to address.

© Chris Chapman and Kerry Rodden 2023
C. Chapman and K. Rodden, *Quantitative User Experience Research*,
https://doi.org/10.1007/978-1-4842-9268-6_8

None of those is a bad goal, but we believe the best way to address any of them is to have a simple, coherent approach to listen to customers. This requires a combination of appropriate sampling, quantitative and qualitative data analysis, and a consistent approach over time.

We advocate simple customer satisfaction surveys that include both a satisfaction rating and an open-ended comment field. A minimal example is shown in Figure 8-1. As we will describe, any satisfaction project should collect both numeric and qualitative information similar to this example.

1. How satisfied are you with [product]?

   o Very Satisfied
   o Satisfied
   o Neither Satisfied nor Dissatisfied
   o Dissatisfied
   o Very Dissatisfied

2. Why did you say you are [piped answer, e.g., satisfied] with [product]? _____

Figure 8-1. *A minimal customer satisfaction survey*

Does the survey in Figure 8-1 seem very short? Well, it is—one of our key points in this chapter is that short surveys are highly advantageous.

8.2 The Components of Listening to Customers

At a minimum, a CSat program needs the following:

- A defined customer population of interest, from which you collect responses (a "sample")

- A survey or other data collection mechanism that will obtain:

  - Ordinal ratings of customers' satisfaction with your product

  - Open-ended comments that describe what customers like or dislike about your product

  - Basic demographic information—although, as we will see, these are not necessarily important for the reasons that stakeholders expect

- A way to iterate and compare samples over time

- Follow-up with stakeholders and customers

Let's examine each of those in turn. We'll describe common questions and problems, and common choices to answer them.

If you're already highly familiar with CSat projects, you might skip ahead to Section 8.3, where we discuss common problems and then present an example analysis in R. Still, we suggest that you skim the following sections, as we hope that our opinions will spur reflection (and perhaps even agreement).

8.2.1 Customer Population and Sample

Before starting, a CSat program should carefully consider which customers to assess and how to reach them. You might assume that *all* customers should be assessed, or at least some random sample of them. Unfortunately, that is impossible due to survey *non-response bias*, the fact that many or most customers do not respond to surveys.

Making the issue worse, those who do respond are likely to differ from a random sample. Customers respond more often when they are fans of a product or, alternatively, have specific complaints.

Another question is whether the business is more interested in some customers than others. It may make business sense to give more attention to paying customers, large customers, accounts that are growing, or those with some other characteristic.

The most common answer to these questions is to attempt to assess *all* customers, not to weight them by importance, and to treat whatever data you collect—regardless of non-response bias—as "the data." In short, gather whatever data you can, and treat that as the best available data. However, if you do this, we urge you to remain aware of the problem of non-response and to collect demographic data that will help you determine whether your sample is representative.

If you choose to collect "whatever data we can get," it is also crucially important to collect future data in exactly the same way, from exactly the same sources. We say more about this and show an example of how you might otherwise draw incorrect conclusions in Sections 8.2.6 and 8.4.

8.2.2 Survey Mechanism

There are three common sources for CSat samples. One is an *email survey* to customers. This assumes that you are only interested in customers for whom you have email addresses. Email surveys are also subject to technical issues such as email filtering. Bulk email providers are often filtered, and such filtering may interact with customers' preferences and demographics (e.g., whether they are heavy email users) as well as industry, organization, and technology platforms. In other words, email survey responses are not independent of customer characteristics and cannot be assumed to be random samples.

However, a great advantage of email surveys is that you can be relatively confident that the respondents are real users because you pulled their email address from your database. Also—with appropriate attention to privacy—you may be able to combine their survey responses with other account information. This can shorten the survey for customers and allow more powerful analyses.

A negative aspect of email surveys is that they are not *blind*, meaning that the recipient will know that you sent the survey. A branded survey can be useful. For example, it lets customers know that the survey is legitimate and that you want to hear from them. However, if you are interested in a competitive assessment—how your product compares to others—an email survey from you may receive biased responses.

A second approach is a survey conducted (or at least recruited) by a third-party provider who reaches respondents in various ways. Depending on the provider, industry, need, and location, this may include email surveys, web surveys, telephone surveys, and in-person interviews or intercepts. A database of respondents used by such providers is called a *panel*. Panel members may be used for multiple studies or sampled repeatedly over time.

The primary problem with an external sample is that you won't know who is an actual customer or potential customer. You have to ask respondents, and their answers are often unreliable and inflated. That leads to the necessity of *screening items* to determine a respondent's interest or status. For example, if your target customers are programmers, you could add a question that tests programming knowledge. However, screening items can be unreliable. You should not assume that you can simply write them and trust that they work as intended. Screening items should be validated to know that they reach their intended audience while rejecting others.

On the positive side, the panel approach is very flexible, can reach potential new customers you can't reach by email, and offers the possibility to do sponsor-neutral ("blind") surveys that are suitable for competitive assessment. There are many panel providers ranging from large, general market research companies to those who specialize in particular audiences such as physicians, IT professionals, or non-English-speaking customers.

A third approach is an *in-product* survey that is delivered while a customer uses your application or service; you have no doubt seen these surveys on websites. The largest advantage of an in-product survey is that it reaches customers immediately, while they are using your product. The context for the survey items is clear and current. These surveys also allow an opportunity (if privacy concerns are satisfied) to tailor the items to particular behaviors or account characteristics, to reach particular customer groups based on that information, and to join the survey data with account information.

On the downside, such surveys only reach currently *active* customers, and they are highly intrusive or even annoying. In effect, you are asking your customer momentarily to stop being a customer because your survey is more important than what they are doing! This may be a reasonable choice if you have a large user base and only need to sample a small proportion of it.

One of the earliest and largest in-product survey platforms was developed at Google while we both worked there. This system is known as Happiness Tracking Surveys (HaTS) and is used by many Google products to conduct short contextual surveys [95]. HaTS results are supplemented by both email and panel surveys (as just described) to gain the advantages and insights offered by each mode.

8.2.3 Ordinal Ratings

The nearly universal survey response option in CSat programs is an ordinal rating scale that ranges from "satisfied" to "dissatisfied" with several levels of intensity. Most commonly, the rating uses a 5-point scale or a 7-point scale, although 4-, 6-, 9-, 10-, 11-, and 101-point scales are not unusual. An example of a 5-point scale is shown in Figure 8-1, where the rating points are

- Very Satisfied

- Satisfied

- Neither Satisfied nor Dissatisfied

- Dissatisfied

- Very Dissatisfied

Even in a single survey item, analysts confront many questions. The following issues are often debated by survey authors:

- The number of points to use, especially 5 points vs. 7 points

- Appropriate adverb modifiers (such as "very satisfied" vs. "extremely satisfied")

- Whether to include a middle point that is neutral

- Whether to label every point or only the end points (for example, to put "very satisfied" at one end and "very dissatisfied" at the other, and not to label anything between them)

- Whether to randomly reverse the scale for half of respondents to avoid order bias

- Whether such a scale may be treated as "continuous" instead of ordinal

- Whether it is better to use average ratings or some sort of proportional metric, such as the proportion who chose one of the top 2 responses (e.g., "satisfied" or "very satisfied")

Our opinions are the following. For the *number of points*, we prefer to include a middle point because many respondents prefer to use it. Either 5-point or 7-point scales are fine, and few respondents are confused by them. Between them, we give the edge to 5-point scales for simplicity, flexibility to fit onto smaller displays, and some evidence that they may work better [113].

Intensifying *adverb labels* ("very," "strongly," "extremely," etc.) should be chosen to match the number of points and what you believe as being appropriate for your customers. Pre-test them to make sure. With a 7-point scale, you might increase the intensity from "somewhat satisfied" to "satisfied" and then "extremely satisfied." A 5-point scale may be better expressed as "satisfied" plus a more intense level ("very," "strongly," "completely," etc.). With 5- or 7-point scales, we label each point to avoid confusion.

Whether to randomly *reverse scales*—that is, to sometimes show the response scale's "best" direction at the top or right hand side, and sometimes instead to show it on the bottom or left hand side—depends on many factors. Depending on the UI layout and cultural context, it may make sense to have a uniform direction for every respondent.

In any case, never alternate the direction *within* a survey for any single respondent, if you have multiple rating items. If direction is randomized, all rating items should be in the same direction within any particular respondent's survey.

Another question is whether responses on a 5-point, 7-point, or other labeled scale may be treated as *continuous* values In this context, the term "continuous" means that analyses regard the ordered response data as falling on an equally spaced metric scale that is convertible to sequential integer values such as 1, 2, 3, 4, 5. The assumption is that the values may be added, averaged, and so forth, which simplifies statistical analysis.

In general, we believe this assumption is unsupportable. There is no reason to believe that there is the same psychological distance between "dissatisfied" and "neither satisfied nor dissatisfied" in Figure 8-1 as there is, for instance, between "satisfied" and "very satisfied." This assumption is driven more by a desire to use simple analytics such as t-tests than by any principled reason. We recommend to use ordinal statistics instead, or convert the ratings to proportions as we describe next. At the same time, experience shows that treating ordinal data as continuous tends to have relatively minor consequences on the results and their interpretation, so we don't object strongly.

8.2.3.1 Top 2 Box and Proportional Scores

Reducing a rating scale to proportions—such as the percentage of respondents who answered "satisfied" or "very satisfied"—can be very useful. Here's an example. Suppose we ran a survey and observed the following frequencies of response:

Response	Proportion
Very Satisfied	36%
Satisfied	45%
Neither Satisfied nor Dissatisfied	11%
Dissatisfied	4%
Very Dissatisfied	4%

The proportion who said they were *satisfied* or *very satisfied* is 36% + 45% = 81%.

A combined proportion has several advantages: it eliminates the question of metric distance between response options, it is easy to explain to stakeholders, and it focuses attention on the business interpretation, namely, how many respondents say they are at least "satisfied."

This is often known as a *top 2 box* score, the percent of respondents who answered with either the highest or second-highest rating. If you use more than 5 points on your scale, you might want to have a top 3 box score, or something similar that fits your interpretive purpose. We find it much easier to tell business leaders that "81% of respondents say they are satisfied" than to explain whatever it means to have an "average satisfaction of 4.05."

Our primary caution in using top 2 box and similar scores is that you should also report the frequencies of the lower boxes. Although it does not affect the top 2 score, there is a large difference between customers answering "neither" as opposed to "very dissatisfied."

8.2.3.2 What About Net Promoter Scores?

Many firms use the Net Promoter Score (NPS) approach. In Section 4.3 we already discussed some of unrealistic expectations that have been voiced about NPS. In this section, we take a closer look at concerns about NPS.

An NPS score is found in the following way:

1. Ask customers how likely they are to *recommend* your product to others.

2. Collect this on an 11-point (0–10) scale.

3. Find the proportion of "promoters" who answer 9 or 10.

4. Find the proportion of "detractors" who answer 0, 1, 2, 3, 4, 5, or 6.

5. Subtract the proportion of detractors from the proportion of promoters and report the net difference. For example, if 50% of respondents rate 9 or 10, and 30% rate 0–6, then the final score would be 50% promoters – 30% detractors = 20% NPS.

NPS achieved great popularity after being promoted as "the one number you need to grow" [111]. Some proponents claimed that NPS is more highly correlated with firms' future growth than any other metric (see Section 4.3).

There are many concerns with NPS. Among them are the following:

- Products vary widely in whether they are recommended at all. Consumers may recommend restaurants, albums, and plumbers but they rarely have opportunity to recommend an electric utility, operating system upgrade, or brand of sugar.

- There is no reason to believe that a score of 9 or 10 means that one is a "promoter," nor that scores of 4, 5, or 6 (in the middle of the scale) indicate being a "detractor."

- NPS ignores responses from anyone who answered 7 or 8 (these don't count toward the proportions used in calculating NPS).

- Scale usage is known to vary widely by culture. Respondents in some cultures are much less likely to rate 10 for *any* product than those in other cultures (see Section 8.4.6).

- NPS relies on difference scores between proportions. This amplifies small and potentially insignificant changes.

- Empirical evidence contradicts claims that NPS is uniquely associated with future growth [67, 123].

- Many stakeholders expect to compare scores across industries and products, which is inappropriate. We discuss this in Section 8.2.6 .

On the other hand, NPS scores tend to be highly correlated with CSat scores, so the practical differences between them may be small.

Our general feeling is that using NPS invites many more problems than it solves, and a standard CSat score is preferable. In a situation where an executive demands an NPS score and will not budge from that demand, we might go ahead and use it. (To be honest, we also try not to work with stakeholders who devalue our expertise and wish to dictate methods.)

8.2.4 Open-Ended Comments

A numeric CSat rating is of limited value in itself. Designers, engineers, PMs, and stakeholders will want to know *why?* Open-ended survey items can help answer that question.

At a minimum, we suggest that you ask respondents why they gave a particular rating. Question 2 in Figure 8-1 presents an example. You may decide that other wording is more appropriate, such as asking, "What could we improve?" However, we like the direct approach of simply asking why they gave their rating. That question is neutral and avoids the presumption that there must be something to improve.

A common problem with open-ended questions is asking too many of them. Most respondents will not say much, and repeated questions risk annoying them, causing them to discontinue the survey, repeat answers, or respond with nonsense. We prefer to ask as few open-ended questions as are necessary, and that is typically only one or two questions. In addition to the "why?" question as shown in Figure 8-1, we often end a longer survey with a wide-open question, "Is there anything else you'd like to say about [product] or this survey?"

How does one analyze open-ended comments? That is beyond the scope of this book, although we identify four progressively more complex approaches that may be employed:

1. *Foundational* (always do): Read all the comments. Break them out by positive and negative ratings, and look for themes.

2. *Consider*: Train a few *coders* to rate comments according to thematic elements (such as particular product features).

3. *Text analytics option*: Apply natural language sentiment analysis, to find key words or themes.

4. *Advanced text analytics*: Perform competitive analysis of satisfaction and comments across your product and competition. Use any of the above methods along with statistical models of product positioning.

It is important to pair survey analytics with qualitative research involving actual users who give feedback in a usability lab or their own environment. Inevitably, you will want to know much more than respondents say in their short responses to a survey item. Qualitative field work will add those details.

Generally, we find that a combination of qualitative work plus approach 1 in the list—simply reading the comments and extracting themes—is sufficient for most projects. Over time, you may wish to add more advanced analytics as in approach 2, 3, or 4. In Section 8.6 we suggest ways to start with sentiment analysis and text analytics for competing products.

8.2.5 Demographic and Behavioral Information

It is helpful to collect some amount of *demographic* information from respondents—although we also urge caution for reasons we'll describe. Demographic information might include a respondent's location, age, income, level of education, household composition, and so forth. You may wish also to collect *behavioral* information such as feature usage, users' goals, competing product usage, and so forth. Demographic and behavior information may come directly from survey items, or from other systems such as customer management systems or product instrumentation.

The usual reason to collect such information is to slice the data by it. Which customers are most satisfied? Least satisfied? What can we do to improve the product for them?

Another important reason is to ensure the *representation* of your sample. You want samples that reflect both your broad user base and (often) the larger population. Unless you collect demographic information, you will not know whether your sample is representative.

As we mentioned in Section 8.2.1, these data are helpful to evaluate non-response bias. If some particular demographic groups are over- or under-responding, you may be able to target data collection by working with a sample provider or other data source. It may even be possible to use these data to reweight, balance, stratify, or match your sample so that it better approximates the population of interest. Those statistical techniques are out of scope here, but we point to references for them in Section 8.6.

Warning: Despite all the good reasons to collect demographic data, there are three large concerns to consider carefully.

First, collecting demographic data makes the survey longer. Every additional survey item increases the risk that a respondent will discontinue (drop out), get tired, start answering randomly or unreliably, or become annoyed. Respondents are not survey-taking robots. Only ask as much as is necessary.

Second, data such as age, gender identification, ethnic identity, and so forth may be sensitive, may annoy respondents, and collecting such information may incur legal obligations for data protection, depending on your location and research locations. Check with your relevant legal counsel or research ethics specialist about those issues. In general, don't collect such data unless you have both organizational permission and an actual need beyond simple exploratory analysis.

Third, there is a specific *analysis* risk that we often observe: slicing data by demographics, observing differences, and then speculating post hoc about the reasons for those differences. The problem is that such analysis invites rampant and nearly unavoidable stereotyping.

Here's an example (disguised, although based on some actual cases). Imagine you have data from US consumers, and it reveals that Latino respondents are especially interested in product usage for school and education. Stakeholders who lack experience with this community may be tempted to invent reasons that reflect stereotypes, such as "They want to learn English" or "Education is a gateway to success for immigrants." Yet those are imagined explanations that have nothing to do with the direct data you have observed.

A simpler explanation—based on real data rather than speculation—is that Latino respondents tend to be younger on average [107] and therefore are more likely than older demographic groups to have children currently at home [81]. So the relationship in our imaginary data could reflect the rather obvious association between *children at home* and *interest in school*, because of an average difference in ages, with little or nothing to do with ethnicity.

The unfortunate reality is that slicing data by such factors invites problematic speculation. What should you do instead? Focus on demographic data to ensure *representation* and do not use it for post hoc analysis. Avoid exploratory data analysis based on demographic factors, especially sensitive ones such as race, ethnicity, or gender. Instead, examine differences related to those factors only when there is a predetermined reason, such as ensuring equity of experience for underrepresented groups.

Do not present data sliced by such factors as part of general baseline reports. If you do uncover an exploratory difference, consider how to explore it more thoroughly and investigate other explanations before drawing conclusions or sharing it broadly.

8.2.6 Don't Compare Groups, Compare Over Time

In Section 8.2.1 we discussed the importance of carefully specifying a target population for assessment. There is a somewhat nonobvious corollary to that: do not compare CSat across different populations, such as different countries, spoken languages, or disparate customer bases. From a technical point of view, when samples come from different populations, most inferential statistical tests are meaningless. "Is there a difference?" Yes, because the data come from different populations!

Different groups use rating scales in different ways. For example, professional users may judge a product more critically than consumers; and some cultures tend to be more effusive in praise than others. In some cultures, a rating below 4 stars is relatively "bad" whereas in other cultures a rating of 3 stars may be viewed as quite satisfactory. In Section 8.4 we will demonstrate how an analyst may be misled by the results of an analysis when this fact is ignored.

The solution to this problem is to compare CSat *within a population, over time.* CSat should be viewed primarily as a longitudinal assessment, to determine whether your product's ratings are changing over time. A good practice is to begin such a project with an initial data collection period to determine the baseline level of satisfaction, and to adapt the data collection methods (survey, response scales, panel sources, etc.).

Once the baseline period has passed, do not change the sample sources, rating questions, or rating scale. Those should remain consistent over time to ensure comparability to past data.

8.2.7 Follow-up with Stakeholders and Customers

When the analyses are complete, you will want to report the results to stakeholders. We note in this discussion some of the most important things to include in a report (Section 8.2.7.1). We also encourage you to consider how the results may be used to "close the loop" with customers (Section 8.2.7.2).

You will want to consider a few aspects of the stakeholder relationship before launching a CSat project. First, clarify the expectations with stakeholders, and pay particular attention to any of the common problems that appear (see Section 8.3). Be especially wary of benchmarks set without prior data (such as, "we want 90% satisfaction"). Try to diminish any expectation that a survey could be used to detect change, such as the impact of a feature release.

Second, be cautious about committing to a "product-like" delivery of results, such as a dashboard or automation into a data warehouse. The problem with those commitments is that initial enthusiasm will turn into a long-term demand for support, even if key stakeholders lose interest. We always say that we would prefer to do new research than to be dashboard engineers. This is not to discount or dismiss the importance of reporting mechanisms. Rather, you should consider carefully whether maintenance of a dashboard or data warehouse would be of interest to you or maximize your value to the organization.

8.2.7.1 What to Report

For CSat projects, you should report the following:

- Sample characteristics including sample size, time frame, the target population(s), and demographics to demonstrate that it is a reasonable sample

- Raw proportions for CSat observations, preferably with confidence intervals

- (Optional) Top 2 box proportions with confidence intervals

- Any change in the above metrics over time, relative to the confidence intervals

- Analysis of any key differences by market or demographics (with caution, as noted in Section 8.2.5)

- Whenever possible, an appropriate competitive product of interest (taking care with regard to potential sample differences)

- Qualitative assessment of the comments that explain any of the observations, level of satisfaction, or reasons for change in observed satisfaction

- Recommendations for the product and business

When it comes to reporting *confidence intervals* we encourage you to avoid discussion of statistical significance and similar technical issues. Instead, it can be effective to visualize the data with confidence intervals, and interpret that chart rather than discussing statistical coefficients, p-values, or the like. In Section 8.4, we demonstrate relevant statistical analyses and charts using R.

8.2.7.2 Closing the Loop

Another question to consider is whether to link a CSat program with direct customer engagement, known as "closed-loop feedback" or "closing the loop" (CTL). A firm might reach out to dissatisfied customers to address problems, offer refunds or discounts, or otherwise engage and improve their experiences. This question most commonly arises for high-value products with repeating customer bases (such as business services, frequent flier programs, luxury goods, and so forth).

Although it is always good to follow up and address customers' problems, a CSat effort is not necessarily a good way to do that. It may make more sense to have a robust customer support program while using CSat assessment to examine performance of the overall product line or services. In that way, CTL and CSat efforts can be individually optimized and will complement one another rather than being tied together. However, there are cases where direct linkage may be appropriate, such as high-value products with small customer bases.

8.3 Common Problems in CSat Analysis

There are several common problems with CSat projects that we've observed over the years. There is not enough space here to describe or address every one of them in detail, but we want you to be aware of them.

- *Unrealistic goals*: Stakeholders often expect a CSat effort to yield deep insight, detect momentary change, or demonstrate that customers are very highly satisfied. None of those is easy or realistic, at least without substantial investigation.

- *Long surveys*: Most surveys are far too long, asking things that "might be important" with no clear relationship to any decision at stake. If you want to hear from customers, ask them what *they* think and keep your own data collection to a minimum. We have often fielded CSat surveys with as few as two or three questions (as in the example shown earlier in Figure 8-1).

- *Unchanging ratings*: It is common for CSat to remain constant for an extended time period. Designers, PMs, and other stakeholders may be perplexed by this and ask, "We released great feature X! Why isn't CSat going up?" There are many possible answers to this, but two fundamental ones are that customers don't care as much as you do about your features; and respondents often answer such questions with a high-level brand evaluation, which changes slowly. Also, when a product is performing well, an *unchanging* rating is a good thing. The important thing is to use CSat as an indicator of health.

- *Comparing populations and products*: Stakeholders often want to compare satisfaction between user groups, countries, product lines, or their own products vs. competitors. We described this problem in Section 8.2.6. Different products are used by different populations, and those populations may respond with different *baseline* scale usage that makes them impossible to compare. Avoid such comparisons as much as possible, and look instead to track CSat over time.

- *Cultural differences in scale usage*: A particularly common example of the previous problem is comparing CSat across cultures who may express themselves quite differently on satisfaction scales. In our experience, it is common to observe high scores in the United States, India, and Brazil, for example, and relatively lower scores in Germany and Japan. That is not because customers are intrinsically more or less satisfied in any particular country but because the cultural expressions and expectations are different. Again, assess *within* a population over time.

- *Driver analysis*: Stakeholders often want to know which factors lead to or "drive" CSat. This is often approached with a large survey and a regression model to look for predictors of CSat; see Chapter 7 in [25] or in [127]. Unfortunately, such data often has high *collinearity* (correlation among items) that makes statistical modeling difficult or impossible. We largely recommend to avoid such efforts and to focus on qualitative assessment of reasons for satisfaction or dissatisfaction. If you decide to pursue driver analysis, you'll want to investigate collinearity and dimensional reduction (discussed in Section 8.6). This involves iterative research to understand relationships in the data and build models that isolate and estimate the important effects.

- *Executive compensation tied to CSat*: Some organizations tie compensation to satisfaction goals, such as X% salary increase if CSat goes up by Y%. This leads to attention to CSat for the wrong reasons (individual gain) and pressures analysts to avoid bad news because their leaders may not want to hear it. In short, the goal to make executives accountable is well intentioned but often turns out badly. Rather than setting goals based on CSat, we encourage organizations to set goals around having a robust listening and CSat *program*, to set goals for how many customers they *reach* and (separately from CSat) *close the loop* (see Section 8.2.7.2), and generally to set goals around learning from customers and maximizing engagement with them.

8.4 Example Analysis in R

In this section, we walk through basic analytics and data visualization for example CSat data. We use the R language [109] as described in Section 6.1.

Our goal here is to show some of the steps in a realistic analysis. This includes data visualization, progressive refinement of analyses and charts, and seeing how deeper investigation may change your initial interpretation.

You should load the data and follow along. The code in this section will load the data for you if you have an Internet connection; or see Section 1.6 for other options. You may find the code at the book's website `https://quantuxbook.com`, or—a good way to learn—type it into R as you read.

If you are new to the R language but program in another language, this code should be generally understandable to you. The specific programming skills required needed for these analyses are covered by Chapters 1–7 in the R companion book [25], and similarly for the Python language in the Python companion, Chapters 1–7 [127].

8.4.1 Initial Data Inspection

In this book's data sets, the file `csat-data.csv` provides simulated data with three variables: the date of response (`Date`), an ordinal rating of satisfaction on a 5-point scale (`Rating`), and the location of the respondent (`Country`). It does not include qualitative comments because text analysis is beyond our scope here (see Section 8.6).

We load the data and check its overall structure (see the book's website for other options to download the data):

```
csat.data <- read.csv("https://quantuxbook.com/data/csat-data.csv")
str(csat.data)
```

```
## 'data.frame':   36048 obs. of  3 variables:
## $ Date    : chr "2020-01-01" "2020-01-01" "2020-01-01" "2020-01-01" ...
## $ Rating  : int 5 1 5 4 5 5 4 5 5 4 ...

## $ Country : chr "US" "US" "DE" "DE" ...
```

This shows that Date and Country are both character (text) strings. By setting those to date and factor types, we can help R apply appropriate statistical models and make it easier to check data quality. We set Rating to be an ordinal (ordered) type rather than a numeric integer (see Section 8.2.3). This protects us from accidentally treating the ratings as interval scale data in later analyses.

We set the data types, using the lubridate package for dates [55], and summarize the data set:

```
# set data types
library(lubridate)      # install if needed
csat.data$Date    <- as_date(csat.data$Date)
csat.data$Country <- factor(csat.data$Country)
csat.data$Rating  <- ordered(csat.data$Rating)

# basic data check
summary(csat.data)

##       Date              Rating       Country
##   Min.   :2020-01-01   1: 1803    DE: 9781
##   1st Qu.:2020-06-30   2: 1786    US:26267
##   Median :2020-12-31   3: 3584
##   Mean   :2020-12-29   4:14464
##   3rd Qu.:2021-07-01   5:14411
##   Max.   :2021-12-30
```

We see that the data were collected for 2 years, the entirety of 2020 and 2021. The observations range appropriately from 1–5 with no outliers or erroneous responses. The respondents were in two locations, the United States ("US") and Germany ("DE"), with roughly 73% in the United States.

This all looks good and is suitable for analysis. In practice, we would do additional data checking [25], but we skip those here for brevity.

8.4.2 CSat for One Time Period

We are always interested to report the latest observations. Let's suppose that those are one month of data. The following code selects a single month, October 2021, from the 2 years of data:

```
csat.month <- subset(csat.data,
                     Date >= "2021-10-01" & Date <= "2021-10-31")
summary(csat.month)

##       Date             Rating    Country
##  Min.   :2021-10-01   1: 59     DE:554
##  1st Qu.:2021-10-08   2: 77     US:934
##  Median :2021-10-17   3:155
##  Mean   :2021-10-16   4:598
##  3rd Qu.:2021-10-25   5:599
##  Max.   :2021-10-31
```

The summary shows us that we selected the right Date range and have roughly the number of other observations as expected.

Next we plot the proportions of observations using ggplot2 [147], noting that proportions are generally easier to interpret than raw counts. We use the scales package [151] to display the proportions as percentages. The result is shown in Figure 8-2.

```
# plot ratings
library(ggplot2)
library(scales)
ggplot(aes(x=Rating), data=csat.month) +
  geom_bar(aes(y=(..count..)/sum(..count..))) +
  scale_y_continuous(labels=percent_format()) +
  xlab("Rating on 1-5 Scale") +
  ylab("Percent of Users Giving Rating") +
  ggtitle(paste0("Satisfaction Ratings, Oct 2021 (N=",
                 nrow(csat.month), ")"))
```

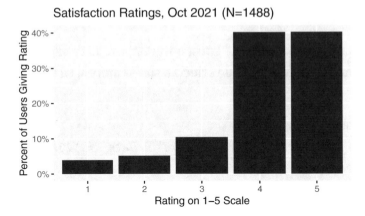

Figure 8-2. *Proportions of CSat responses in one month of data*

In the ggplot command, we added `geom_bar()` with code to calculate the proportions, and used `scale_y_continuous()` to have the axis `labels` display percentages (instead of raw proportions, 0–1).

This chart is not bad, but it would be better to add confidence intervals (CIs). In an actual project, we might find confidence intervals using an ordinal estimation or a bootstrapping process, although we will leave that aside for now (see Section 8.6). A simpler approximation—which is adequate for charting and directional interpretation—is to treat each observation as a binomial observation (either *this* rating or *another* rating) and use binomial confidence intervals as estimates of multinomial CIs.

We calculate the 95% binomial CIs using the traditional normal approximation formula, where *p* is the proportion and *N* is the sample size, multiplying by 1.96 units of standard error to obtain the 95% estimate:

$$CI = \pm 1.96 \sqrt{\frac{p*(1-p)}{N}}$$

See Chapman and Feit [25], Section 6.3.2, for details on binomial confidence intervals and alternative estimation options.

A proportion chart with CIs is so useful that we'll make a reusable *function* for it, shown next. That makes it easy to produce the same format for other data sets (see Section 8.4.6) and it puts intermediate calculations out of the way rather than cluttering up our primary memory space in R.

```
plot.csat.ci  <- function(dat, titleDate="") {
  csat.month.ci <- data.frame(table(dat$Rating)) # Frequencies
  names(csat.month.ci) <- c("Rating", "Freq")
  csat.month.ci$Prop  <- csat.month.ci$Freq / sum(csat.month.ci$Freq)
  csat.month.ci$N     <- sum(csat.month.ci$Freq)
  # then compute 95% CI using binomial normal approximation
  csat.month.ci$ci    <- sqrt(csat.month.ci$Prop * (1-csat.month.ci$Prop) /
                               (csat.month.ci$N)) * 1.96
  csat.month.ci$ciLo <- csat.month.ci$Prop - csat.month.ci$ci
  csat.month.ci$ciHi <- csat.month.ci$Prop + csat.month.ci$ci

  # plot histogram with CI
  ggplot(aes(x=Rating, y=Prop, ymin=ciLo, ymax=ciHi),
            data=csat.month.ci) +
    geom_col(fill="gray") +
    geom_errorbar(width=0.2, color="darkred") +
    scale_y_continuous(labels=percent_format()) +
    coord_cartesian(ylim=c(0, 0.45)) +
    xlab("Rating on 1-5 Scale") +
    ylab("% Giving Rating (with 95% CI)") +
    ggtitle(paste0("CSat Ratings, ", titleDate,
                   " (N=", nrow(dat), ")")) +
    theme_minimal()
}
```

In this function, we first find the proportions for each rating, and then calculate CIs according to the binomial formula. Finally, we plot those CIs and return the plot object (which by default will also be drawn in the R plot window). In the plotting routine, error bars are added with geom_errorbar(). The option theme_minimal() gives a cleaner visual style.

A single line of code will produce the plot:

```
plot.csat.ci(csat.month, "October 2021")
```

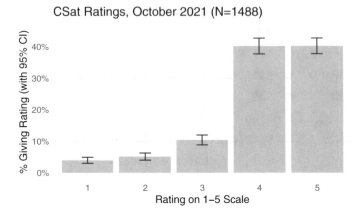

Figure 8-3. *One month of CSat data with confidence intervals*

This improved chart with confidence intervals is shown in Figure 8-3. It reports N=1488 observations, which our function found by looking at the number of rows in the data we passed into it (nrow(dat)). For simplicity, we directly specified the date to put in the title; that might also have been determined by looking at the data. We can now read the confidence intervals from the chart. For the rating of 1, we see in the chart that the CI ranges from approximately 3.0%–5.0%, and we may read the ranges similarly for other ratings.

Figure 8-3 is exactly the kind of chart we routinely include in CSat reports. It includes sample metadata (date and sample size) ensuring that the chart could be interpreted reasonably well on its own, apart from descriptions. Does the reusable plot.csat.ci() function appear complex? Perhaps it is, at first; but as we'll see later, it will accelerate our work.

8.4.3 CSat over Time

The next important thing to report is whether CSat is changing over time (see Section 8.2.6). For an initial examination, we plot ratings by Date, as shown in Figure 8-4. To do this, we momentarily treat the ratings as continuous integer values using the as. numeric() function. Section 8.2.3 describes why this is not a great idea, and we'll look at alternatives in a moment. However, it is useful as a quick check at the beginning of an analysis.

In this case, we assign the chart to an object p that will hold the chart so we can adjust it later:

```
# Next let's look at all 2 years of data
# What if we want the average, and treat ratings as simple numeric values
(p <- ggplot(aes(x=Date, y=as.numeric(Rating)), data=csat.data) +
  stat_smooth() +
  scale_x_date(date_breaks = "3 months", date_labels="%b %Y",
               date_minor_breaks = "1 month", expand=c(0, 1)) +
  theme(axis.text.x = element_text(size=7)) +
  ylab("Average satisfaction (5 point scale)") )
```

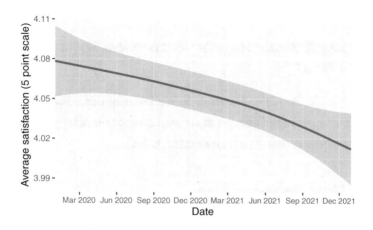

Figure 8-4. *Mean rating and confidence interval over time, treating the ratings as integer (continuous) values*

The resulting chart is shown in Figure 8-4. Notice that the stat_smooth() function automatically adds a shaded band for the confidence intervals while plotting the average value as a solid line. Unfortunately, a problem with the chart is that the Y axis is restricted to a narrow range, which visually overemphasizes the amount of change. Because we saved the chart as object p, it is easy to adjust the range with additional code to expand the Y axis:

```
# fix Y axis
p + coord_cartesian(ylim=c(3.0, 5.0))
```

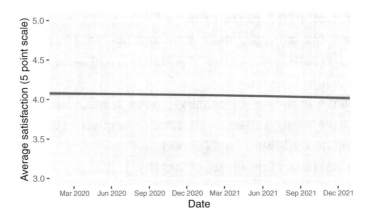

Figure 8-5. *The longitudinal change in CSat is placed into a clearer context with an adjusted Y axis range*

The updated chart is shown in Figure 8-5. We see that there is a small but gradual decline in the numeric average CSat. Next we will look at plotting proportions instead of this arguably inappropriate average (see Section 8.2.3).

8.4.4 Top 2 Box Proportions

As previously discussed, top 2 box scores are common and are advantageous for simplicity in interpretation (Section 8.2.3). To calculate a top 2 box proportion, create a new variable that assigns each observation to either the top box (above some cutoff value) or not. For these data, the top 2 boxes are values of 4 and 5 in our ordered `Rating` variable, so we assign responses to the top 2 box (or not) as follows:

```
# Better: treat individual ratings as proportions in themselves
# where 100 == "in top 2 box", 0 == "not in top 2"
csat.data$Proportion <- ifelse(csat.data$Rating >= 4, 1, 0)
```

In this code, the new variable `Proportion` gets a value of 1 when the ordinal `Rating` falls into the top 2 boxes (values of 4 or 5), and otherwise gets a value of 0 when an observation is not in those top 2 boxes.

Using `stat_smooth()` we plot the average values and confidence intervals for that newly created variable `Proportion`. That plots the top 2 box score by date, as shown in Figure 8-6.

```
ggplot(aes(x=Date, y=Proportion), data=csat.data) +
  stat_smooth() +
  coord_cartesian(ylim=c(0.7, 0.9)) +
  scale_x_date(date_breaks = "3 months", date_labels="%b %Y",
               date_minor_breaks = "1 month", expand=c(0, 1)) +
  scale_y_continuous(labels=percent_format()) +
  theme(axis.text.x = element_text(size=7)) +
  ylab("Top 2 Box %") +
  ggtitle("Satisfaction trend (Top 2 Box %)")
```

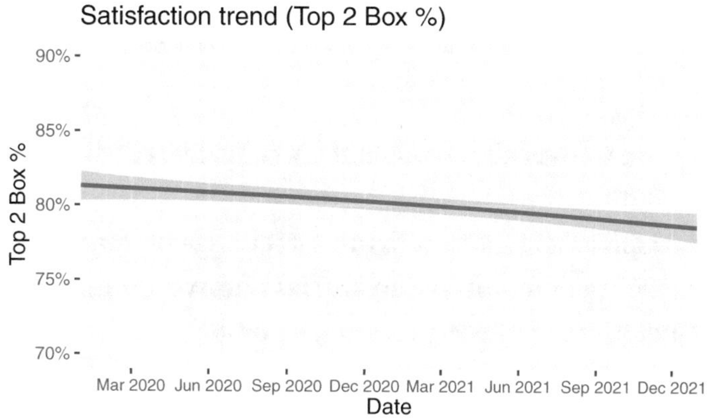

Figure 8-6. *Top 2 box proportion by date*

Over the 2-year observation period, the top 2 box proportion in these data (simulated data) decreased from approximately 81% to 78%.

That is not a large decline, although it is certainly undesirable. If these were real data, we would expect stakeholders to be concerned. How concerned should we be? What else might be occurring in these data? We'll investigate those questions in the next sections.

8.4.5 Is CSat Changing? Initial Analysis

Figure 8-6 shows an apparent decrease in CSat over time. As a first analysis, we might ask whether there is a constant rate of change over time that differs from zero (i.e., differs from *no change*). We estimate a simple linear model for change in the proportion of satisfied respondents over time as follows:

```
lm1 <- lm(Proportion ~ Date, data=csat.data)
summary(lm1)

## Call:
## lm(formula = Proportion ~ Date, data = csat.data)
...
## Coefficients:
##               Estimate Std. Error t value Pr(>|t|)
## (Intercept)  1.544e+00  1.851e-01   8.341  < 2e-16 ***
## Date        -3.989e-05  9.937e-06  -4.014 5.98e-05 ***
## ---
## Signif. codes:  0 '***' 0.001 '**' 0.01 '*' 0.05 '.' 0.1 ' ' 1
...
```

This model shows a statistically significant and negative effect of Date—a decrease over time.

The estimated coefficient for Date is in units of change in *proportion* per *day* (a decrease of –0.00003989 per day). To get an annual rate, we multiply that by 365 days, and also by 100 to make it readable as percentage points:

```
coef(lm1)["Date"] * 365 * 100 # est'd change in Top2 % for 1 year

##     Date
## -1.455864
```

This says that that the top 2 box score declined by 1.46 percentage points per year in these data. There is an important caveat: that is the change *according to this simple linear model*.

But is that the best model? Remember that we also have data for respondents' locations, given in the Country variable. Let's see whether that affects our interpretation.

8.4.6 Examination by Country

First, let's summarize the data by country. We use an *anonymous function* to calculate proportions for each rating:

```
aggregate(Rating ~ Country, data=csat.data,
          function(x) prop.table(table(x)) * 100)
```

```
##   Country  Rating.1  Rating.2  Rating.3  Rating.4  Rating.5
## 1      DE 10.868009  8.465392 13.475105 41.284122 25.907371
## 2      US  2.817223  3.647162  8.626794 39.692390 45.216431
```

We find that there are comparatively more ratings of 5 in the United States (45.2%), and more ratings of 1 in Germany (10.9%).

Is the difference statistically significant? A chi-square test checks for the conditional independence of factor data:

```
chisq.test(table(csat.data$Country, csat.data$Rating))
```

```
##  Pearson's Chi-squared test
##
## data:  table(csat.data$Country, csat.data$Rating)
## X-squared = 2095.2, df = 4, p-value < 2.2e-16
```

The low p-value in this result says that the distribution of ratings differs in the two countries.

Let's plot the data by country to see the difference. Remember the function we created earlier to plot a month of CSat data with confidence intervals? It is exactly what we need to compare the United States and Germany side by side; we just add the R subset command to select separate sets of data for each country:

```
plot.csat.ci(subset(csat.data, Country=="US"), "US")
plot.csat.ci(subset(csat.data, Country=="DE"), "Germany")
```

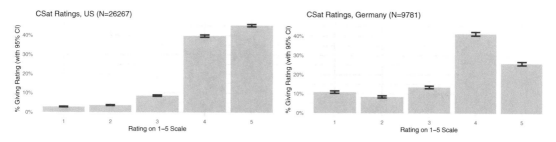

Figure 8-7. *Rating proportions by country*

Figure 8-7 shows the side-by-side charts. The ratings are substantially higher in the United States than in Germany, especially for the proportion of 5 ratings.

This tells us that our model should not consider only the possibility of change by Date as in our initial linear regression model. It should also take into account the fact that ratings differ by Country. We'll do that in the next section.

8.4.7 A Better Model of CSat Change in These Data

We saw that the results differed between the United States and Germany. Now we will update the chart for CSat over time to separate the two countries.

We can separately plot the top 2 box proportions by Country by instructing ggplot to use different colors (and, in this case, line types) for each country:

```
ggplot(aes(x=Date, y=Proportion,
           color=Country, linetype=Country), data=csat.data) +
  stat_smooth() +
  coord_cartesian(ylim=c(0.50, 0.90)) +
  scale_x_date(date_breaks = "3 months", date_labels="%b %Y",
               date_minor_breaks = "1 month", expand=c(0, 1)) +
  scale_y_continuous(labels=percent_format()) +
  theme(axis.text.x = element_text(size=7)) +
  ylab("Top 2 Box %") +
  ggtitle("Satisfaction trend by Country (Top 2 Box %)")
```

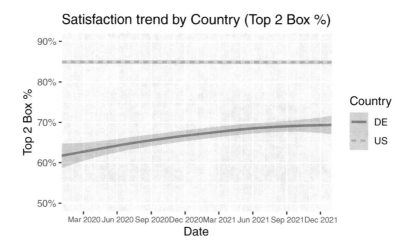

Figure 8-8. *Top 2 box proportion by country*

The result is shown in Figure 8-8, which tells a strikingly different story than Figure 8-6 where the data were pooled for both countries. Figure 8-8 shows that CSat in the United States has remained essentially unchanged over two years, while CSat in Germany has actually *increased*.

We check that with a regression model that includes `Country` as a main effect (reflecting the baseline difference in rating scale usage in the two countries) while adding an interaction effect between `Country` and `Date` (to examine whether the United States and Germany show differential rates of change over time). In R:

```
# A better linear model includes country and change for country by time
lm2 <- lm(Proportion ~ Country + Date:Country, data=csat.data)
summary(lm2)

## Call:
## lm(formula = Proportion ~ Country + Date:Country, data = csat.data)
...
## Coefficients:
##                   Estimate Std. Error t value Pr(>|t|)
## (Intercept)      -1.364e+00  3.638e-01  -3.749 0.000178 ***
## CountryUS         2.312e+00  4.218e-01   5.482 4.23e-08 ***
## CountryDE:Date    1.090e-04  1.947e-05   5.597 2.20e-08 ***
## CountryUS:Date   -5.338e-06  1.147e-05  -0.465 0.641713
## ---
## Signif. codes:  0 '***' 0.001 '**' 0.01 '*' 0.05 '.' 0.1 ' ' 1
...
```

The interaction coefficient for `CountryUS:Date` (line 4 in the regression coefficients summary) shows the change in the United States over time. The change is very small (–5.338e-06) and the standard error is larger than the change (1.147e-05), so it is not statistically different from zero in this sample. In other words, there is no substantial change over time in the United States. Meanwhile, CSat is increasing significantly in Germany (`CountryDE:Date` is both positive and much larger than its standard error).

Comparing Figure 8-8 and Figure 8-6, how is this possible? How can the same data show that CSat is steady in the United States and is increasing in Germany…and yet it is declining overall when the two countries are combined? The answer is that the German respondents gave lower ratings, while the *sample proportion* of German respondents grew over time. The lower scale usage there drags down the overall average, and that effect increases over time as the sample from Germany grows larger.

Out of curiosity we can code country membership and use charting code that is nearly identical to how we previously visualized top 2 box proportions:

```
csat.data$USprop <- ifelse(csat.data$Country=="US", 1, 0)
csat.data$DEprop <- ifelse(csat.data$Country=="DE", 1, 0)
```

We visualize the proportions of Country in the sample as follows:

```
ggplot(aes(x=Date, y=USprop, linetype="US"), data=csat.data) +
  stat_smooth() +
  stat_smooth(aes(x=Date, y=DEprop, linetype="DE")) +
  scale_x_date(date_breaks = "3 months", date_labels="%b %Y",
               date_minor_breaks = "1 month", expand=c(0, 1)) +
  scale_y_continuous(labels=percent_format()) +
  theme(axis.text.x = element_text(size=7)) +
  ylab("% respondents, by country") +
  labs(linetype="Country") +
  ggtitle("Sample % over time, by Country")
```

The result is Figure 8-9, which shows the change over time in the relative proportions of respondents by country.

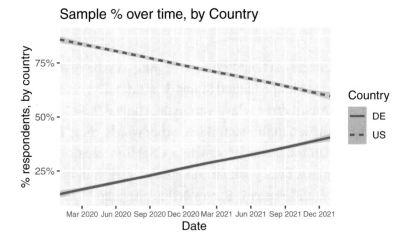

Figure 8-9. *The proportion of the total sample represented by observations from the United States and Germany, over time*

If these were real data, a potential business interpretation of the CSat change (as shown earlier in Figure 8-8) might be this: the customer base grew in Germany and satisfaction improved there, while high satisfaction was maintained in the United States. That is quite a different message than the initial impression of an overall decline in satisfaction!

Is that the correct interpretation? To know for sure, we would need to bring in qualitative data and check whether it agrees. In any case, a naive message that CSat is declining would be incomplete if not outright misleading.

These data are a dramatic example of how cultural differences may affect data. It is not infrequent for a new product to start in the United States, achieve high satisfaction, and then expand to other markets where satisfaction is lower. The crucial thing for Quant UXRs to determine—in partnership with qualitative research—is whether satisfaction is "really" lower, as might be indicated in differences among open-ended and qualitative comments, or whether it reflects a baseline effect of differences in scale usage by culture.

The effect we saw here is known to statisticians as *Simpson's paradox*. In Simpson's paradox, an outcome appears to be different in aggregated data than it appears when constituent groups are analyzed separately. In this case, an apparent overall decline in CSat was not supported by examination of individual countries. Additional examples are described in the companion books for R (Section 9.2.8) and Python (Section 8.2.6). Several additional examples of Simpson's paradox are given by Wang et al [146].

8.5 Key Points

In this chapter, we reviewed best practices for planning, fielding, and analyzing customer satisfaction surveys. Although such surveys appear to be straightforward, there are many ways they go awry in real world practice. The following points will help you to field better and more useful surveys.

- CSat data may be collected in a product or outside of a product, and by using either known customer lists or general audience survey panels. Each channel is valuable yet has limitations. We recommend using multiple channels and data sources when possible (Section 8.2.2).

- Top-line metrics should be used primarily to engage stakeholders to learn more. It is crucial to ask users for qualitative, open-ended feedback and to read, analyze, and report what they have to say (Section 8.2.4).

- Common errors include censored data, non-response bias, negative response bias, cultural differences in scale usage, and asking too many questions (Sections 8.2.1, 8.3).

- It is easy to misinterpret CSat data when the underlying sample dynamics are changing. Keep sampling as consistent as possible (Sections 8.2.6, 8.4). Even when the nominal sample mechanism (such as the survey method) is consistent, a changing mix of respondents may lead to incorrect conclusions.

- Always compare CSat to a baseline such as results from the same sample source in a previous time period. Don't compare across disparate populations such as substantially different products, user groups, or countries (Sections 8.2.6, 8.4.7).

- Top 2 box metrics are common. They have the advantage of being more interpretable than a rating scale, with the disadvantage of losing information (Section 8.2.3.1).

- We don't recommend *difference* scores, such as calculating the top 2 proportion minus the bottom 1 proportion. Net Promoter Score is an example. This approach tends to be noisy and obscure the underlying trends even more than top 2 box proportions (Section 8.2.3.2).

- Net Promoter Score (NPS) is a popular approach, but it suffers from the instability of difference scores, information loss of middle-point raters, questions about cultural applicability, and somewhat excessive expectations (Section 8.2.3.2).

- When your data include multiple groups, take care to examine whether patterns are consistent across them. Be especially careful when analyzing groups that have different patterns. You may need to separate the analyses, include grouping variables in a statistical model, or take other steps to understand trends more clearly (Section 8.4.7).

8.6 Learning More

Bootstrapping. Ratings observations are best treated as *ordinal* data, and a good way to find confidence intervals for such data is to use bootstrapping techniques. Those are too complex to cover in this book, so we point you to a comprehensive overview of the methods by DiCiccio and Efron [40]. That paper is followed by several commentaries in the same issue of *Statistical Science*. A moderately mathematical and readable guide (for a statistics text) that covers such methods in depth and uses them as the basis for a systematic approach to statistical reasoning is Chihara and Hesterberg, *Mathematical Statistics with Resampling and R* [31].

Driver analysis. Driver analysis is a common request from stakeholders who want to know which features, behaviors, or customer characteristics are "causing" satisfaction or dissatisfaction. The problem of assessing causation is extremely difficult, and with rare exceptions, you should expect only to find *correlational* associations rather than causal ones. For more on driver analysis, see Chapter 7 in the R [25] or Python [127] companion books. More generally, texts on linear modeling are applicable; an excellent overview and applied text is Harrell's *Regression Modeling Strategies* [57].

Collinearity. A common problem with satisfaction data is when many variables are highly correlated. This occurs especially often when satisfaction is assessed for multiple aspects of a product, such as asking, "How satisfied are you with your phone? ... with its size? ... its display? ... its memory? ... its battery life?" Customers tend to be satisfied or dissatisfied with most aspects simultaneously. When such data is put into a multivariable statistical model, it leads to a problem known as *collinearity*. In a nutshell, when such ratings are nearly identical, a statistical model cannot determine the differential importance of any one aspect vs. another. To address this, you will need to remediate the collinearity in some way. Strategies to do this are discussed in the R companion, Section 9.1 [25], and Python companion, Section 8.1 [127].

Sample consistency. Analysts often obtain samples whose characteristics are substantially different than those of the larger population of interest. Two common approaches to addressing this are *stratified sampling* and *matched sampling*. These are complex topics, but here is a rather simplified overview. Stratified sampling attempts to control in advance for this problem by collecting data from known strata (groups with different characteristics), and later applies statistical methods that control for the strata differences. Matched sampling adopts more of a retrospective approach, looking at response patterns and then (as just one approach) resampling or weighting the data to be more representative. One way to do that is by finding "matching" respondents from different groups who are otherwise quite similar; there are more sophisticated approaches.

If you are frequently concerned with group differences and whether your data for those groups are representative of a population, we encourage you to learn more about stratified sampling and weighting. A good overview with a moderate dose of undergraduate-level mathematics is Lohr, *Sampling: Design and Analysis* [82].

If you are interested in both sample representation and causal inference, then you might start by investigating matched sampling. An excellent introduction to causal inference using matched sampling, is Cunningham, *Causal Inference: The Mixtape* [38].

Text analytics. Finally, CSat analysis is often combined with *text analytics* to assess customer sentiment, find patterns in open-ended responses, and to uncover topics across comments. A short overview of the questions and methods used for customer text analytics was published by Chris [21] and is available freely online. A complete introduction to applied text analytics is Kwartler, *Text Mining in Practice with R* [77].

8.7 Exercises

These exercises are intended to do three things: to have you write and field a survey, analyze its data, and think about the qualitative aspects in addition to quantitative analysis.

Do you need a topic for a survey? You could choose a product, hobby, family event, nonprofit group, whatever it may be. In general, anyone can write a survey about anything; you could ask about satisfaction with a product even if it's not your own product (although if you're an industry researcher asking about a competing product, there may be limitations; check with your research colleagues or product counsel). If you need ideas, here's a list to consider: smartphones, laptops, an automobile brand, a local restaurant, your Internet provider, the local police department, traffic in your area, commercial air travel, a social media platform, or a coffee chain.

1. For something of interest to you, write a customer satisfaction survey. Be sure to include at least one open-ended comment field. (Google Forms is an easy platform to write simple surveys.) Ask four to eight friends or family members to take the survey and reflect aloud on how they understand the questions. What changes would they suggest? Do you agree?

2. Field the survey online (or create a similar one for something else). Post it to Facebook or some other online forum, and collect responses. Analyze them as shown in this chapter. Are the reasons for satisfaction or dissatisfaction clear from the open-ended comments?

3. Now that you have data, what is it that you *wish* you had asked differently or in addition? Are you certain that your respondents were the right audience? Did they have the experience needed to answer? Update your survey to address some of those issues and field it again.

4. *Advanced.* Additional exercises for the R companion book [25] are freely available at `https://r-marketing.r-forge.r-project.org/exercises/ChapmanFeit2e-Exercises-Only.pdf`. Work through the exercises in Section 7.9 (pp. 15–16 in the Exercises PDF).

CHAPTER 9

Log Sequence Visualization

Applications, websites, electronic devices, and many other products and services yield data detailing *sequences* of user behaviors. For a website, a user sequence may comprise pages visited or actions taken on those pages. For an application, a user sequence may include all of the actions taken while a user engages with the app. As an example, a gaming app might log a player's avatar name, the level at which they started, the movements they made while playing, the various game objects and other players they interacted with, the resulting points scored, and so forth.

With behavioral sequence data, an analyst may approach many questions, including

- Which sequences are most common?

- Which are least common, perhaps unexpectedly so?

- Which user activities commonly occur together and might benefit from careful attention to their joint design?

- Which actions rarely occur together, perhaps indicating a problem or an opportunity for better design?

- How do the sequences relate to user characteristics or product use cases?

- How are users starting and ending their visits to the product?

- Where are users abandoning an important product flow, such as a path to purchase or sign up?

- Do we see changes in behavior as the result of an intervention, such as an A/B experiment, marketing campaign, or new product release?

© Chris Chapman and Kerry Rodden 2023
C. Chapman and K. Rodden, *Quantitative User Experience Research,*

Those questions mostly involve descriptive statistics and basic hypothesis testing. In the R companion book [25], Chapter 14 discusses basic handling of behavioral data, including *sessionizing* the data (grouping the events into *sessions*, according to proximity in time for one user) and descriptive statistics. It then demonstrates how to use Markov chain analysis to look at event probabilities.

In the present chapter, we extend those analyses and explore a popular method to visualize sequence data: *sunburst* diagrams. As you will see, a sunburst visualization can be used to show progressive sequences of actions, including the sequence length, order of actions, and relative proportions of behaviors and events within a sample. Such charts are helpful for descriptive reporting and interactive exploration.

This chapter is primarily a hands-on chapter using R code. First we create example data and show how a sunburst diagram represents a sequence of events. Then we turn to a more realistic (and messy) public data set and visualize its data. To follow along, use R and step through the code as presented in this chapter.

As always, both the code file and data sets are available at the book's website, `https://quantuxbook.com`. The code in this chapter will also load the public data set directly from the Internet. We recommend *typing* the code from the snippets here, for maximal learning.

9.1 Example Sequence Data

We recommend to begin an analysis project with simulated data whenever possible. By working with such data, you can debug your methods and prepare code that will work later when real data arrives. We'll do that now before tackling a real data set in Section 9.2.

For simulated data, we're going to consider what data might look like if we were to log users' events at a breakfast buffet. In this simulation, we track five food items: pastries, granola, yogurt, eggs, and potatoes. For each person, we observe choices of one to four items per buffet trip. A particular person might take only granola; or they might choose a pastry, eggs, and potatoes; or they might get two pastries and yogurt; or four different kinds of granola; or any other combination. (This is not a particularly *realistic* set of buffet data, but that doesn't matter. The goal is to understand how the method works before moving to a real, complex data set.)

We generate these data randomly in R. First we set a random number seed to make the process repeatable. Then we specify the five foods and how many observations we want (5000):

```
set.seed(10010)         # make the data repeatable
foods <- c("Pastry", "Granola", "Yogurt", "Potatoes", "Eggs")
N.obs <- 5000           # we'll simulate 5000 observations
```

For each person, we draw 1–4 foods from the list. A sequence could be longer than four behaviors, but we'll limit it here for simplicity. We use the R `sample()` function to draw the number of foods for each person, and then look at a frequency table of how often we will draw sequences with one food, two foods, and so forth:

```
num.events <- sample(4, N.obs, replace = TRUE)
table(num.events)

## num.events
##    1    2    3    4
## 1244 1223 1307 1226
```

The table shows that 1244 respondents (24.88%) will have a sequence with only one item, while similar proportions of simulated users will have sequences with two, three, or four items.

Next we need to select *which* foods will go into each sequence. We write a function that takes as input the number of foods (1–4) to be drawn and returns a vector (set of values) of randomly selected foods that comprise a sequence of that length. Valid observations, for example, might be "Eggs" (1 item), "Granola-Yogurt" (2 items), "Eggs-Yogurt-Eggs-Granola" (4 items with 1 repeated), and so forth.

The function takes `len` as the number of items and `dat` as the list of possible behaviors (the `foods` set). The parameter `prob` is used to weight the items so they are not chosen with strict uniform probability. The option `replace=TRUE` allows an item to be chosen more than once. Finally, the separator "-" is used to separate the items in a sequence.

That function is shorter to write than to explain! It is as follows:

```
one.event <- function(len, dat, prob=((length(dat)+2):3),
                      replace=TRUE, sep="-") {
  event <- sample(dat, len, prob=prob, replace=replace)
```

```
  paste0(event, collapse=sep)
}
```

As always, we test it for a few cases:

```
one.event(3, foods)
```

```
## [1] "Granola-Yogurt-Pastry"
one.event(6, foods)
```

```
## [1] "Pastry-Potatoes-Yogurt-Potatoes-Yogurt-Potatoes"
```

The first case has three items in the sequence, "Granola-Yogurt-Pastry," while the second case has six items. The latter is more than we'll use in our data but demonstrates that the function works and handles longer sequences with repetition.

Now that we have a function to simulate a sequence for one person, we may apply it to all 5000 cases in the num.events vector by using the sapply() function:

```
events <- sapply(num.events, one.event, dat=foods)
head(events)
```

```
## [1] "Yogurt"                "Granola-Yogurt-Granola" "Pastry"
## [4] "Yogurt"                "Eggs"                   "Potatoes"
```

The events object has 5000 observations, where the first sequence is simply "Yogurt", the second is "Granola-Yogurt-Granola", and so forth. This code is also an example of how functions (see Section 6.2.1) and vectorization (see Section 6.2.2.1) make code shorter and simpler.

9.1.1 Sunburst Chart for the Buffet Data

A sunburst diagram plots the progressive *frequencies* (either counts or proportions) of items that occur in sequences. To get the frequencies, we use the R command table() to count unique occurrences and store the results in a data frame. We do that and then check a few rows (head()), using knitr:kable() for formatted output:

```
events.freq <- data.frame(table(events))
knitr::kable(head(events.freq))
```

events	Freq
Eggs	147
Eggs-Eggs	15
Eggs-Eggs-Eggs	3
Eggs-Eggs-Eggs-Granola	1
Eggs-Eggs-Eggs-Pastry	1
Eggs-Eggs-Eggs-Yogurt	1

The resulting table is alphabetized. The sequence "Eggs" occurs 147 times among our 5000 simulated observations, while "Eggs-Eggs" occurs 15 times, and "Eggs-Eggs-Eggs" occurs 3 times. (Those three people would really love eggs, perhaps choosing an omelet, scrambled eggs, and a hard-boiled egg.)

We use the R sunburstR package [8] to chart the sequence frequencies, adding a color palette from the RColorBrewer package [97]:

```
# May first need to run install.packages("sunburstR")
library(sunburstR)
library(RColorBrewer)    # install if necessary
breakfastPalette <- brewer.pal(5, "Set1")
sunburst(events.freq, colors=breakfastPalette)
```

The result is shown in Figure 9-1. In a sunburst diagram, the inner ring is a *donut chart* (which fits nicely with our breakfast example) showing how often each behavior occurs *first* in a sequence. In this case, "Eggs" is in the upper right area. If you hover the mouse pointer over the chart—which may be done live in R—it shows that 12.4% of sequences begin with "Eggs".

☐ Legend

Figure 9-1. *A sunburst diagram for simulated choice sequences among customers at a breakfast buffet*

Each successive ring after the first one shows an additional step in the observed sequences. Moving upward from "Eggs" in Figure 9-1, we see in the second layer that "Pastry" and "Granola" are the most common second behaviors. The sequences "Eggs-Pastry" and "Eggs-Granola" have overall proportional frequencies of 2.60% and 2.48%, respectively. We could similarly read sequences with 3 or 4 behaviors. At each step, roughly 1/4 of the sequences end, while the other 3/4 continue and add another action.

An example is shown in Figure 9-2, where we hover over "Pastry" and then "Eggs" and finally "Potatoes". That three-item sequence occurs in 0.4% of the observations.

By inspecting a sunburst diagram, we are able to explore event frequencies, common paths across several actions, and see where particular sequences come to an end. This is helpful both to understand users' actions and to uncover places where they have difficulties, or discontinue usage of a product, website, or service. Consider how this might work in an e-commerce application. If we observe one particular path that includes the behavior *add to cart* and yet frequently ends without reaching *checkout*, it would indicate that many users are failing to complete purchases. We would investigate the possible reasons in depth. That might include system issues, user interface confusion, or other problems.

We encourage you to tinker with the code presented in this section. Change the set of behaviors (the list of foods) to a domain that interests you. Try changing the size of that list as well as the length of sequences, and the probability of each item being chosen (via the prob parameter to one.event). Explore the resulting sunburst diagrams.

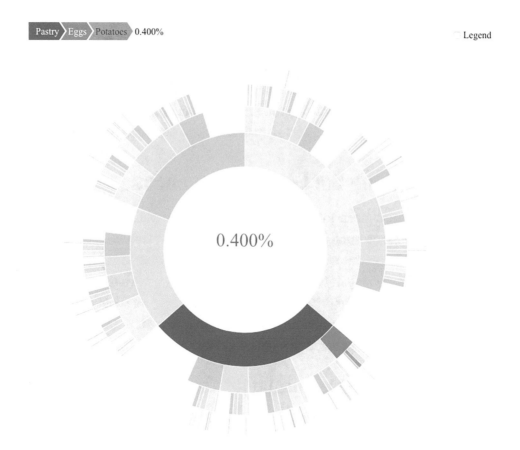

Figure 9-2. *A sunburst diagram shows proportional frequencies for a sequence with multiple events when one hovers over an event in an outer ring. In this case, we highlight the sequence "Pastry-Eggs-Potatoes" in the lower right direction*

The next section builds on this foundation and examines a real data set of website visits.

9.2 Sunburst Visualization of Website Data

Thomas Edison is reported to have said that his inventions depended on "one percent inspiration and ninety-nine percent perspiration" [98]. Similarly, data analysis projects often depend on ninety-nine percent data handling and one percent inspired analysis. In this section, we take a real data set and wrangle it into a format suitable for the R package, `sunburstR`. After formatting the data, it requires minimal code to draw a sunburst diagram.

The data here are a single day from the anonymous logs of a web server operated by the United States Environmental Protection Agency (US EPA) in the 1990s [9]. The logs contain each request made to the site (such as loading an HTML page or image) along with a time stamp, the originating Internet address (IP address), and other data. In the R companion book, these data were used for basic log and session statistics, along with Markov chain modeling of users' navigation [25]. In this chapter, we go beyond those analyses to visualize users' page request sequences with a sunburst diagram.

9.2.1 Transforming the Logs to Sequences

As you saw in the previous section, the `sunburstR` package expects sequences to be separated with hyphens, such as "`event1-event2-event3`". However, web and application logs are almost always sequential lists with several or many *rows* per user. Those data require processing in two steps. First, a user may use a site or application in several discrete time periods for different tasks and their sequences need to be divided into *sessions* based on time gaps. Second, once the sessions are identified, the events must be aggregated into the format that `sunburstR` expects.

9.2.1.1 Loading and Sessionizing the Data

In this section, we load the EPA web server data and divide it into sessions. This is described in detail in Chapter 14 of the R companion book [25]. For our purposes here, we give the complete code with minimal explanation.

First, we load the data from a saved R object and view the first few rows:

```
# web log data after basic processing (Chapman & Feit, Ch. 14)
epa.df <- readRDS(gzcon(url("https://goo.gl/s5vjWz")))
knitr::kable(head(epa.df[ , 1:3]))
```

host	timestamp	request
host195	[29:23:53:25]	GET /Software.html
host1888	[29:23:53:36]	GET /Consumer.html
host2120	[29:23:53:53]	GET /News.html
host2273	[29:23:54:15]	GET /
host2273	[29:23:54:16]	GET /icons/circle_logo_small.gif
host2273	[29:23:54:18]	GET /logos/small_gopher.gif

The data compiles sequential events of requests to get HTML pages, GIF images, and the like. Each host is an anonymized IP address, corresponding roughly to a user.

To prepare to divide the data into sessions, we first sort it by datetime within each unique user (host), and then calculate the time in minutes between each successive observation:

```
# 1. put DF in order of host and timestamp
epa.ordered <- epa.df[order(epa.df$host, epa.df$datetime), ]

# 2. get time differences between rows in minutes
epa.ordered$time.diff <-
  c(NA, as.numeric(
          epa.ordered$datetime[2:nrow(epa.ordered)] -
          epa.ordered$datetime[1:(nrow(epa.ordered)-1)],
        units="mins"))
```

When there is a large gap in time, we assume that the data fall into a new session. More specifically, a new session is indicated by either (a) a new user or (b) a gap of 15 minutes or longer within one user. We flag those conditions as applicable for each row of the data:

```
# 3. determine new sessions, as being either:
# .. 1: host has changed since previous row
# .. 2: time difference exceeds session cutoff time of 15 minutes
session.time              <- 15   # exceed (mins) ==> new session
epa.ordered$newsession    <- NA   # is this row a new session?
epa.ordered$newsession[1] <- TRUE # row 1 is always a new session
```

```
# session is new if host differs from previous row
# or the max session time is exceeded
epa.ordered$newsession[2:nrow(epa.ordered)] <-
  ifelse(epa.ordered$host[2:nrow(epa.ordered)] !=
          epa.ordered$host[1:(nrow(epa.ordered)-1)],
        TRUE,
        epa.ordered$time.diff[2:nrow(epa.ordered)] >=
          session.time)
```

The newsession column is TRUE for the first row of every unique session.

Finally, we assign sequential session numbers, and remove all the events except for HTML page views. If we were interested in other behaviors, such as loading image files, we could leave those requests in the file. However, a web page often automatically requests many image files, which show up as individual events in a log file. Their presence makes sequences very long, and doesn't tell us anything new about the user's browsing behavior. So here we limit our analysis to the HTML page events:

```
# 4. finalize session numbers & initial time differences
epa.ordered$session <- cumsum(epa.ordered$newsession)
epa.ordered$time.diff[epa.ordered$newsession] <- NA # NA if new

# 5. select only the HTML events
epa.html <- epa.ordered[epa.ordered$pagetype=="html", ]
```

We always check results along the way:

```
epa.html[1:5, c(1, 13, 10)]
```

```
##              host session                                  page
## 12383  host10         2                 /docs/ozone/index.html
## 13891  host10         3                          /Research.html
## 13928  host10         3               /docs/campus/campus.html
## 14248  host10         3                      /docs/BioTech.html
## 12182 host100         4 /docs/OPPTS_Harmonized/abguide.txt.html
```

In this case, the first session (numbered "2") shows a request for a single page (ozone/index.html). The next session browsed three pages (Research.html, campus. html, and BioTech.html).

To understand basic data such as page view frequencies, the number of unique users, or the number of sessions per user, we could apply the R `table()` function and other descriptive statistic methods to these values in `epa.html`. Those analyses are described in the R companion [25], Chapter 14. We skip them here and focus on preparing the data for sunburst visualization.

9.2.1.2 Creating Sequences from the Sessions

These data have been labeled with unique session identifiers (`session`). Now we will assemble them into single line strings that combine the events from each session, as we did previously for the breakfast buffet sequences.

The `sunburstR` package separates items with a hyphen ("-") but that character may also appear in HTML file names. For purposes here, we replace every "-" character in an HTML page request event with an underscore ("_"):

```
# changes hyphens in the page names to underscores (_)
epa.html$page <- gsub("-", "_", epa.html$page)
```

Next, we divide the data frame into individual sessions. This will let us easily work on each one to compile a sequence string.

```
# split epa.html$page into a separate data frame for each session
epa.chunks <- split(epa.html$page, epa.html$session)
epa.chunks[1:2]  # check a couple of the sequences
```

```
## $`2`
## [1] "/docs/ozone/index.html"
##
## $`3`
## [1] "/Research.html"          "/docs/campus/campus.html"
## [3] "/docs/BioTech.html"
```

The last step is to take the individual page requests in each chunk and stick them together, separated by hyphens. This is not difficult in R although the code needs some explanation.

First, here is the code:

```
# assemble events into 1 sequence string for each chunk
# set max length of 5 events, for tidier sunburst
```

```
epa.sequences <- data.frame(sequence=sapply(
  epa.chunks,
  function(x)
    paste0(x[1:min(length(x), 5)], collapse="-")))
```

We will start with the inner part of the code (the last line) and work outward to explain it.

To begin, we limit the sequences to a maximum length of 5 events. That is arbitrary, but in these data there are few sequences longer than that. More generally, this would be decided based on the characteristics of your data and problem. To select anywhere from 1 to 5 events, we specify the vector index as `1:min(length(x), 5)`. When it has only 1–4 items, the `min(length(x), 5)` function guarantees we don't try to select more items than exist (`length()`); but if the sequence is longer than 5, it will stop at 5 (which would be the minimum value when compared to the length).

Next, we paste the 1–5 events into a single sequence string using `paste0()`, putting a hyphen between each pair (`collapse="-"`). This combination of selecting 1–5 items and pasting them together is placed into an *anonymous function* (see Section 2.7.2 in [25]). That function is run on every one of the sequences in `epa.chunks` using the vectorized `sapply()` function. The final result—the set of all the individual sequences—is stored in the `epa.sequences` object.

We inspect the first two (and in practice would check many more) to make sure everything looks good:

```
epa.sequences[1:2, ]
```

```
## [1] "/docs/ozone/index.html"
## [2] "/Research.html-/docs/campus/campus.html-/docs/BioTech.html"
```

As expected the first sequence has a single page view, while the second one comprises three events. This matches what we previously observed when we inspected the sequence chunks (`epa.chunks[1:2]`) and the HTML page sequence (`epa.html[1:5, ]`).

The data set is now ready for sunburst visualization.

9.2.2 Sunburst Visualization of the EPA Data

Sunburst diagrams work with frequencies of sequence occurrence. We obtain those using `table()` and inspect the first three results:

```
# count the occurrences of each sequence
epa.sequences.freq <- data.frame(table(epa.sequences$sequence))
head(epa.sequences.freq, 3)
```

```
##                                                          Var1 Freq
## 1          //Standards.html-//Rules.html-//Rules.html      1
## 2                                      /305b/sum1.html      9
## 3 /305b/sum1.html-/305b/sum1.html-/305b/h2o_mon.html      1
```

The first and third sequences occur only once, while the second one (which is a single page view) occurs nine times.

If we plot every sequence that happens a single time, our diagram will be extremely difficult to read. We need to put a cutoff somewhere. How about sequences that occur more than once (two times or more)? If we limit the chart to those, how many would there be? We use `table()` with a Boolean test to check:

```
# how many sequences occur more than once?
table(epa.sequences.freq$Freq > 1)
```

```
##
## FALSE   TRUE
##  1371    149
```

Our test evaluates to TRUE when a sequence occurs more than once. There are 149 sequences in this data set that occur at least twice, roughly 10% of the total number of unique sequences. We will plot those sequences—that occur two or more times—in our sunburst diagram. (Note: it may be an important finding that 90% of user sequences occur only once! We'll leave that aside for now, and say a bit more in Section 9.4.)

We draw the chart using `subset()` to select the sequences that occur more than once and `sunburst()` to create the diagram:

```
library(sunburstR)
sunburst(data=subset(epa.sequences.freq, Freq > 1))
```

The result is shown in Figure 9-3.

Before exploring individual sequences, several patterns are immediately evident in Figure 9-3. First, there is no single starting point with more than roughly 10% prevalence (7.8% in fact). Rather, there are many different starting points with relatively or very low prevalence rates. This suggests that users are coming from a variety of disparate sources, perhaps search engine results, external links into specific pages, bookmarks they have saved, and so forth.

Next, a majority of the sequences are only a single page. This is consistent with the specificity of the many starting points, and the use of links from other sites to arrive directly at a relevant page, rather than navigating there from the site's home page. Presumably many users find what they want on the single page they visit, although it is also possible that they are referred into a page, find it to be useless, and immediately leave. Deeper analysis—such as time spent on a page, plus qualitative research—could provide insight and confirmation.

In the second ring, several of the second pages are colored identically to the first page in the sequence, which means users are reloading the same page. That may be a signal of user confusion or poor page performance. Those would be good candidates for investigations of page loading times, error rates (such as HTML 404 errors, which are in the log data), or simply observing users and asking them.

Figure 9-3. *Sunburst diagram for the EPA website data, selecting sequences that were observed at least twice*

In a live version of the diagram—such as on a website, a browser, or an RStudio plot window—we can hover over a sequence with the mouse to drill into its steps. Figure 9-4 shows one such sequence, a path from `ozone/index.html` to subpages relating to science, Q&A, and process.

Figure 9-4 shows the kind of sequence that would be of high interest to owners of the "`ozone`" section on the website. They are likely to have questions about why this sequence is relatively more frequent than others; what sequences may be missing that they would hope to see; what users are trying to accomplish on this path; and so forth.

This relatively large set of sequences demonstrates an advantage of sunburst visualizations over alternatives such as flow and Sankey diagrams: sunburst visualizations are able to present hundreds of sequences in a compact display. If you have only a small number of distinct sequences, such as 5 or 10 (perhaps up to 20 or so), then a flow or Sankey diagram [103] may be a better visualization choice.

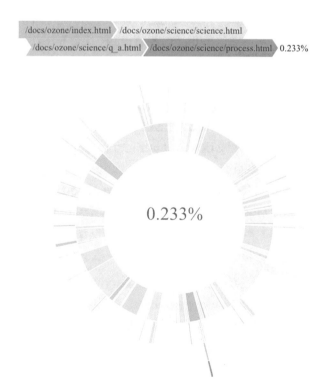

Figure 9-4. *Highlighting a single path with four events in the sunburst diagram. The path shown in the lower right of the chart occurs in 0.233% of observed sequences*

9.2.3 Next Steps in Analysis

To go further than the inferences and questions presented in the previous discussion, we would need to know more about the data set, the website design, and stakeholder questions. Although this is a real data set, it does not come with any context, and data analysis without context is always limited.

With that in mind, we would consider a few next steps:

1. Are there patterns that might benefit from a different kind of analysis? In some systems, users repeatedly cycle between pages, such as viewing a list, taking action on an entry, returning to the list, and repeating. In those cases an analysis using transition frequencies (e.g., Markov chain analysis; see Section 9.4) can be a good supplement to sunburst visualization.

2. Can the pages be grouped into meaningful, higher-order categories? There may be collections of pages that have different URLs but represent very similar user actions or intentions, and could therefore be collapsed into the same category. For example, we could replace the full URL path with a category name such as [search] or [materials]. Such choices are dependent on the goals of the analysis and a detailed understanding of the data. When done well, categorization makes a sunburst diagram clearer and less cluttered [115]. It also results in fewer sequences that occur only once, which we filtered out in this first pass. Ideally, all of those filtered sequences would be placed in a higher-order category so that no data is omitted.

3. What additional *user observation* do we need? Visualization of this sort is primarily an exploratory exercise, and exploration always raises additional questions. A common next step is a targeted usability assessment that looks at particular sequence areas in depth. For instance, suppose we look at the sequence in Figure 9-4 from ozone/index to process and wonder why users take that path. We could talk to UX research colleagues to find out what

they know about this particular user journey. If more information is needed, we could construct a usability lab study that asks users to start at ozone/index and discuss their next steps and thought processes.

4. Are there other kinds of user assessment that would be helpful? A sunburst diagram may suggest places to intercept users and ask them about their satisfaction. For example, the "ozone team" might develop a short survey to ask users whether they are finding the information they want. This survey could pop up on selected pages, either randomly chosen or targeted to pages that the team would like to investigate. The sequence data can help us in two ways: to find appropriate locations for such assessment, and to estimate in advance the number of users we are likely to reach for CSat or other assessment (see Chapter 8, "Customer Satisfaction Surveys").

5. How might our findings inform metric selection? Sequence analysis can uncover evidence of task failure or difficulty, especially when combined with validation and explanation from user studies or surveys. For example, a frequent path of several searches in a row without a click through to a result could indicate a problem with the search results not being relevant enough. Observing such patterns of behavior can help provide ideas for using logs data to create metrics of task success, as discussed in Chapter 7, "Metrics of User Experience."

6. Are there other ways that our analysis might be helpful to engineering teams? Among other things, an engineering team should conduct formal testing of a product's user interface with simulated users. This may include random bots that engage with the product following scripted interaction sequences, among other tests. Our observations of user sequences (and related usability assessment) can inform the test team and help them focus on core user areas, as well as uncover unanticipated paths for them to test. (By the way, be sure to filter out the data from any such automated testing in your analysis!)

7. How best to present the findings? Sunburst visualizations are more suited to data exploration and understanding than to stakeholder presentation or decision-making, and you may need to explain your discoveries using simpler charts. For example, if there is an important single sequence with a pattern of steep user drop-off, this could be more clearly presented as a basic funnel or bar chart.

The upcoming "Learning More" section suggests additional directions for behavioral log analysis.

9.3 Key Points

Following are important points to keep in mind about sunburst visualizations and related sequence analysis methods:

- The data are observational and analysis is primarily *exploratory*. Key findings are usually descriptive and often suggest additional areas for analysis, visualization, or user research (Section 9.2.3).

- Sunburst diagrams present *sequential steps*, such as user actions on a website or in an application. They visualize the successive conditional frequencies of observed steps using nested, interactive donut charts (Section 9.1.1).

- Given the availability of libraries like `sunburstR`, most of the effort in sunburst visualization is taken by data preparation. This includes sessionizing the data, reformatting it, assembling it into sequences, filtering it, and potentially aggregating and anonymizing it (Sections 9.2.1, 9.2.1.2).

- Unless your data include predetermined *sessions*, you will need to determine a principled way to break the user data into chunks that represent sequences of related behaviors. This involves domain knowledge. A common method is to use gaps in time when a user was inactive (Section 9.2.1.1).

- In your first pass at a sunburst visualization, you will probably find that you have too many distinct categories in your data, with a *long tail* of rare sequences or actions. It is worthwhile to create higher-order categories and iterate on the visualization to make it more coherent and useful (Section 9.2.3).

- Potential *next steps* after creating an initial visualization include engaging in additional user observation such as usability lab sessions; assessing specific user paths or end points, perhaps to understand users' goals or satisfaction, or to help define new metrics; and engaging with engineering teams to inform product design and testing (Section 9.2.3).

9.4 Learning More

Sequences sunburst. Kerry originated the application of the sunburst visualization to sequence data, and has written separately about the motivations and implementation, including a case study from YouTube [115]. That article presents code that is not dependent on R and works with JavaScript, using the D3.js library (see also [117]). The R sunburstR package is based on similar D3 code. Sunburst visualizations use hierarchical, or *tree*, data, where a core structure breaks off into multiple branches and leaves. If you work extensively with sequence data and sunburst visualizations, you will want to understand algorithms for such data, as in Sedgewick and Wayne, *Algorithms* [128], Chapter 4. When using a sunburst visualization, it is important to be aware that its radial layout leads to some distortion of the data as you move outward through the rings. See [116] for a detailed discussion of the advantages and disadvantages of radial visualizations.

   Trajectory analysis. A *trajectory* is a set of sequential states over time, and trajectory analysis is often used to analyze changes over individuals' lives, such as their sequences of family behaviors, education, health status, careers, and so forth. More generally, if you have sequence data that are fundamentally related to an implicit (or explicit) standard of *time*, then trajectory analysis may be useful. What do we mean that data is "related to time?" Consider the website sequences in this chapter. Although they occur in time, the analysis of time is of little importance apart from determining session status; we don't really care whether a user takes 10 seconds on a page vs. 1 minute, or whether they visited the EPA site in the first month of their job vs. their 10th year. By contrast, progress through educational levels, through life stages such as marriage or divorce, or through technology behaviors such as purchasing devices or achieving levels in a game does have an important reference to time. With the EPA data, if we wanted to analyze sequences according to the hour of the day, or day of the week, then it would be time-related. This is largely a matter of analyst choice and depends on the questions at hand. When you are concerned with such change over the course of time, the R TraMineR ("trajectory miner") package contains both exploratory and inferential statistical methods for state analysis, including multiple forms of visualization [48].

Markov chain analysis. Sunburst diagrams are primarily an exploratory tool that is a visualization of descriptive statistics (frequencies). A different approach would be to ask a more specifically *parametric* question: given some particular state (a specific behavior), what is likely to happen next? One way to approach that is with Markov chains, which model the likelihood of successive events such as user behaviors actions, conditional on previous actions. Markov chain analysis is a useful and complementary method to sunburst visualization. Chapter 14 in the R companion [25] gives an introduction to Markov chain analysis. An outstanding general introduction is given by Grinstead and Snell, *Introduction to Probability* [54].

9.5 Exercises

These exercises build on the sunburst visualizations presented in this chapter. Several of them ask you to write custom code. Among those, Exercises 3–5 can be completed using the code covered in this chapter (and some minor extensions of it), while Exercise 6 requires other reference sources and a modest degree of coding experience beyond what is shown here, at the level of general fluency in R or Python.

1. *[Thought exercise].* For what other kinds of user sequence data might a sunburst diagram be useful?

2. *[Thought exercise].* Choose some other activity—not necessarily related to "users"—and detail some of the elements that could appear in sequence data. Some ideas include browsing items online or in a bookstore, performing dance moves, hearing ads on a podcast, or doing reps (repetitions) of a physical exercise.

3. *[Code, basic]*. Modify the breakfast buffet example to create sequences for one of the activities you considered in Exercise 1 or 2. Write the code to create the sequences and then visualize them. Adjust the code to make the sequences slightly more realistic (for example, by varying the probability of each item being chosen).

4. *[Code, moderate]*. The LSApp data set (`https://github.com/aliannejadi/LSApp`) contains sequences of mobile app usage for 292 participants [1]. Download the `lsapp.tsv.gz` file and decompress it, which should give you a 179MB file that is tab-delimited and can be read into R with `read.delim()`. Follow the steps in this chapter to create an initial sunburst visualization using this data set. Note that the data contain some long runs where a single app name is repeated many times, and you will need to find a way to collapse these into one instance (hint: the R `rle()` function may help).

5. *[Code, moderate]*. After creating the initial sunburst visualization in Exercise 4, see if you can make it into a more meaningful analysis of app usage. For example, try mapping some or all of the 87 app names to a smaller number of categories (Messaging, Games, etc.) and using those categories as the step names. This will reduce the number of possible combinations, and therefore the number of sequences that appear only once overall, making it possible to avoid having to filter those out of the analysis.

6. *[Code, moderate]*. Section 9.4 briefly described Markov chain analysis, which is another useful method for analyzing user sequences. Review one of the references for Markov chains, and define a Markov chain transition matrix that will simulate user behavioral sequences. Format and visualize the resulting sequences.

CHAPTER 10

MaxDiff: Prioritizing Features and User Needs

The MaxDiff survey method has been rapidly gaining popularity among UX researchers. MaxDiff uses forced-choice trade-off questions to determine the scaled order of user preference for a list of items—such as product features, messages, use cases, blocking issues, or many other kinds of data. The results can be used to inform product strategy, meaning that MaxDiff offers a way for Quant UXRs to apply their skills in the earliest stages of the product lifecycle, when no other user data is available.

In this chapter, we describe common use cases for MaxDiff surveys and why it is often a better approach than traditional stack rank or Likert scale methods. We illustrate MaxDiff with R code, first for a simple data set (pizza preferences), and then in more detail with simulated data of user preferences for seeking data online.

There are two ways to read this chapter. First, you may wish to follow along with the code in R. If you plan to run MaxDiff surveys using a general survey platform such as Qualtrics CoreXM, or you want the most thorough understanding of MaxDiff, we recommend to interact with the R code. That will give you the deepest exposure to MaxDiff models and the underlying format of its data and statistical models. On the other hand, the code in this chapter is more complex than the code in previous chapters and requires a larger and slightly fragile set of packages in R.

The second way to read this chapter is to read it for the *concepts* of MaxDiff and postpone engaging with the R code. This is suitable for initial learning about the method and for readers who will use a specialized platform that supports MaxDiff, such as Sawtooth Software's Lighthouse Studio (discussed in Section 10.2.3). In this case, skip over the code in the latter half of the chapter, and focus instead on the explanations and interpretation of results.

© Chris Chapman and Kerry Rodden 2023
C. Chapman and K. Rodden, *Quantitative User Experience Research*,
https://doi.org/10.1007/978-1-4842-9268-6_10

If you have difficulty with the R code, check carefully that you have all the packages that are required, especially those noted in Section 10.3.4.1. It may help to install the latest version of R [109] and update your installed packages. See Section 1.6.2 for additional assistance with debugging.

As with the other code chapters in this book, both the R code file and data sets are available at `https://quantuxbook.com`. The code here is also suitable to be typed into R and will automatically load data from the website via an Internet connection.

10.1 Overview of MaxDiff

The easiest way to understand MaxDiff is to see an example. Suppose you're starting a restaurant and would like to know which kinds of pizza—among some artisan options— that your customers would prefer. One approach would be to put the pizzas on your menu and see what sells; but that is inefficient and it doesn't help you to know what supplies you should order in advance to have on hand. A faster method would be a preference survey.

As for the specific pizzas, we'll assume you are considering 11 different pizzas and we'll refer to them with short names, such as *Arugula* and *Arrabbiata*. On a survey, you would want to explain what the ingredients are for a pizza with a given name, perhaps with reference to a menu, photos, descriptions on mouse rollover during the survey, or some other guide to the details. We'll assume you have done that and anyone who takes your survey will understand what the pizza names mean and what ingredients they have. For convenience, we'll refer to the short names.

How do you find out which pizzas your customers prefer? One approach would be to ask them to rate the pizzas on a likelihood scale, such as shown in Figure 10-1.

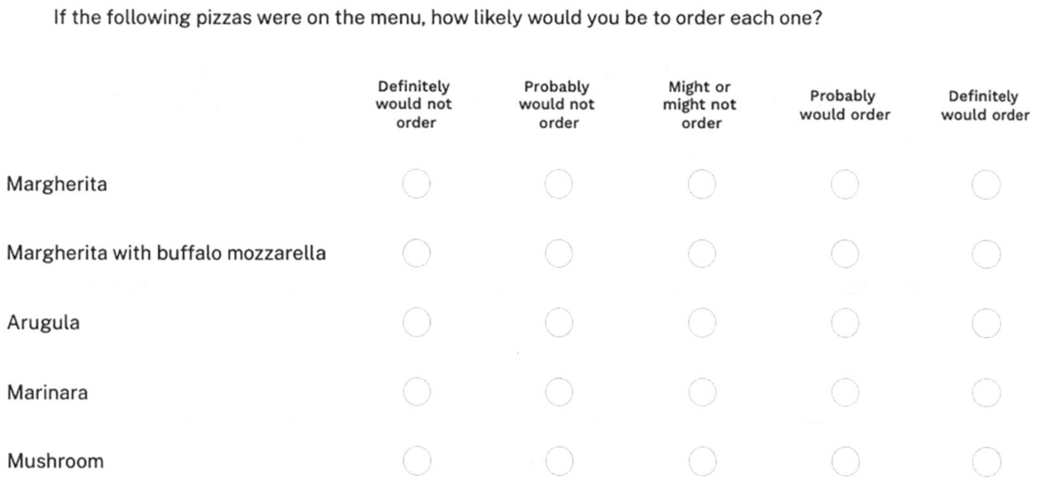

If the following pizzas were on the menu, how likely would you be to order each one?

	Definitely would not order	Probably would not order	Might or might not order	Probably would order	Definitely would order
Margherita	○	○	○	○	○
Margherita with buffalo mozzarella	○	○	○	○	○
Arugula	○	○	○	○	○
Marinara	○	○	○	○	○
Mushroom	○	○	○	○	○

Figure 10-1. *A choice likelihood grid for pizza preference. This is difficult for respondents to answer and for analysts to interpret. A MaxDiff alternative has many advantages because it forces trade-offs among options, so respondents report what they prefer most*

There are several problems with the scale in Figure 10-1:

- The question is difficult for respondents to answer. They are asked to compare too many options at once. What if the menu had 60 pizzas? Can someone answer reliably? (Chris has often eaten at a restaurant in Italy that offers more than 60 pizzas.)

- The answers are ordinal in nature—"definitely" is more than "probably"...but how much more? Is that distance ("4" to "5") exactly the same distance as between "probably not" and "might or might not" ("2" to "3")? How are you supposed to aggregate the answers?

- Closely, related, it does not translate directly to choice likelihood. Suppose a customer says "probably"...is that 51% or 75% or 95% likely? Does "definitely" mean exactly 100%?

- It also does not force trade-offs. Suppose a customer answers "definitely" to three different pizzas, under the view that they would "definitely order one of them". But which one? Or are they saying they would order all three? You don't know.

- • Customers can also straight-line respond. It would be perfectly reasonable (and potentially honest) to answer "might or might not" to every pizza. But that kind of data is not helpful to estimate actual demand.

In short, an ordinal grid of this kind is impossible to translate into what you really want to know, which is the demand for each pizza. For example, suppose that Margherita pizza and Pepperoni score the highest with an average response of "Probably would order." How many would you expect to sell? How many ingredients should you order? There is no way to determine the degree of preference needed to answer those.

At the core, you need to understand the *choices* that your customers will make, but an ordinal scale is far removed from a choice process. MaxDiff items more closely simulate customers' choices and trade-offs. Figure 10-2 shows an example MaxDiff survey question asking about a respondent's pizza preference, expressed as a direct choice.

A MaxDiff item breaks a longer list (in this case, all 11 pizzas) into smaller chunks and asks about only a few items at a time, typically 4 or 5 items. Respondents choose the single "most likely" and the single "least likely" option from each small set of randomly selected items. After that, they see a new random set and again choose the best and worst options. That may be repeated six times, eight times, or more (we'll discuss that below).

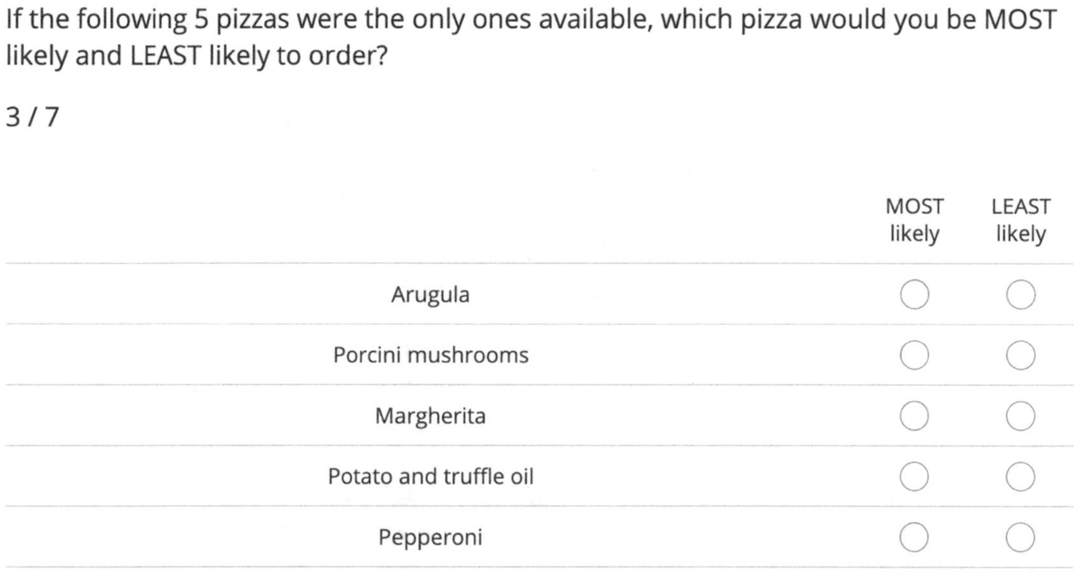

If the following 5 pizzas were the only ones available, which pizza would you be MOST likely and LEAST likely to order?

3 / 7

	MOST likely	LEAST likely
Arugula	○	○
Porcini mushrooms	○	○
Margherita	○	○
Potato and truffle oil	○	○
Pepperoni	○	○

Figure 10-2. A MaxDiff item for pizza choice. The task is limited to a small number of choices and then repeated across additional random sets of choices (the screenshot here indicates that it is number 3 out of 7 total choices)

Importantly, respondents are not able to equivocate or say that all choices are equally likely—they are forced to indicate preference. You might wonder, what if they really have no preference? Well, in that case it doesn't hurt to select one of the items as best or worst, because they'll have additional chances to answer.

How do you move from these observed responses of "most likely" and "least likely" to obtain more specific estimates of the level of preference? The answer is that you apply a statistical model that estimates the probabilities of choice from the observed data. We'll review the details of that model later in the chapter. First, we present an example of the analysis.

10.1.1 Illustration of MaxDiff Analysis

In this section, we will quickly present code and initial results for a small data set on pizza preferences. Our goal in this section is *not* to explain the code—which we will do later in the chapter—but to give you a preview of how MaxDiff works and the kind of results you get.

The data here come from a MaxDiff exercise fielded using Qualtrics CoreXM for a list of 11 pizza varieties, with items as shown in Figure 10-2.

First we load the choicetools library [23], check the data structure, and read the actual data. For the code snippets in this section, some setup is required that we discuss later in the chapter (Section 10.3.4.1). It would be perfectly OK for you just to follow along by reading the code for now, and wait until later in the chapter to try it hands-on. Alternatively, you could install the materials as noted in Section 10.3.4.1 and then return here to continue.

We'll describe these steps in detail later, but for now the code is as follows:

```
library(choicetools)   # choicetools installation details are later
download.file(url = "https://bit.ly/3FvVoNE",
              destfile = "qualtrics-pizza-maxdiff.csv",
              method="auto")
# change the file location to match your system as needed
md.define <- parse.md.qualtrics("qualtrics-pizza-maxdiff.csv",
                                returnList=TRUE)
md.define$md.block <- read.md.qualtrics(md.define)$md.block
```

When this code loads the data, it displays various messages; we omit those for now and will discuss the details later.

Next we estimate a statistical model for the choice probabilities. Again, we omit diagnostic information that is shown as the model runs.

```
test.hb <- md.hb(md.define, mcmc.iters = 10000,
                 mcmc.seed=98101)                       # estimation
md.define$md.model.hb    <- test.hb$md.model            # save results
md.define$md.hb.betas    <- test.hb$md.hb.betas         # raw coefficients
md.define$md.hb.betas.zc <- test.hb$md.hb.betas.zc # individual scores
```

The model gives us the overall averages for the *sample*, known as an *upper-level* model, and for each individual *respondent*, known as the *lower-level* estimates (or just *upper* and *lower models*, for short). We say more about those in Section 10.2.5.3.

The `choicetools` package includes functions to plot the upper-level (sample) estimates and the individual distribution. The function `plot.md.range()` plots the upper-level estimates with Bayesian credible intervals. We plot the ranges, remove some chart clutter (`theme_minimal()`), and add an axis label:

```
plot.md.range(md.define) +
  theme_minimal() + xlab("Pizza Toppings")
```

The results are shown in Figure 10-3. In these data, the most preferred pizza has truffles, followed by a Margherita pizza with buffalo mozzarella. After the top four pizzas, the next five are roughly tied in preference. The Margherita and Pepperoni pizzas are in the far last two places in preference. Does that mean that Margherita and Pepperoni pizzas are unpopular in general? No, only that they are unpopular relative to the other choices, *in this data set*. We'll say more about relative preference, sampling, and sample size later.

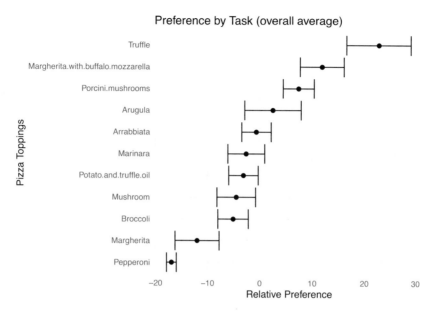

Figure 10-3. *Average, sample-level preferences for pizzas from a MaxDiff exercise with N=6 respondents*

For any aggregate, average results such as those shown in Figure 10-3 you should ask, "What about the individuals? Do they all agree or are there large individual differences?" MaxDiff gives individual-level estimates that help to answer this. Using the `choicetools` package, we plot the distribution of individual preferences as follows:

```
plot.md.indiv(md.define) +
  theme_minimal() + ylab("Pizza Toppings")
```

Figure 10-4 shows a small circle for each individual's response estimates. Examining the top two pizzas, we see that Truffle was very highly preferred and there were no respondents who strongly disliked either the Truffle pizza or the Margherita with buffalo mozzarella. Similarly, the other pizza varieties all had relatively tight sets of preferences and all were broadly acceptable, with the exception of the Margherita and Pepperoni pizzas. For those two pizzas, almost every respondent preferred every other pizza type to either one of them.

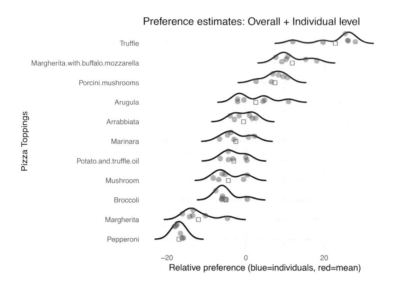

Figure 10-4. *The individual estimates for pizza preference in our N=6 data set*

Are you wondering what's up with these unusual preferences? In the general population, pepperoni pizza would be more popular than this. Here's why we observe these preferences: these data came from Chris answering the survey six times, and he is vegetarian. This is not a general sample (unless you're making pizza for Chris). On the other hand, it shows how well MaxDiff can work with even a small sample—as long as you are clear about what population is represented in the results. In Section 10.2.2.5 we discuss issues for sampling and sample sizes for MaxDiff in more detail.

10.1.2 Calculating Pizza Demand

We posed the question, how should we stock our restaurant according to the best guess of pizza demand? The estimated preferences from MaxDiff can answer that for us. We describe the details in Section 10.2.5.2. As a preview, here's how it works.

First, we choose the set of pizzas we want to compare. Suppose we decide to offer six pizzas on our menu: Margherita, Mushroom, Arrabbiata, Pepperoni, Arugula, and Marinara. Although these are not the most popular in our data (see Figure 10-4), they may be the most desirable for business reasons (supply availability, the cost of truffles, and so forth). We get the estimated preference values for those pizzas:

```
pizza.util <- md.define$md.hb.betas[ ,
          c("Margherita", "Mushroom", "Arrabbiata",
          "Pepperoni", "Arugula", "Marinara")]
```

We use those values in a short equation known as the *multinomial logit* formula (Section 10.2.5.2). That calculates the relative preference share for each pizza, compared to the entire set of six pizzas on the menu. We multiply by 100 to convert proportions to percentages, and round it for easy interpretation:

```
round(100 * exp(colMeans(pizza.util)) /
          sum(exp(colMeans(pizza.util))))
```

```
## Margherita    Mushroom Arrabbiata  Pepperoni    Arugula    Marinara
##           1           9          23          0         51          14
```

Among those six pizzas—given the data here—the Arugula pizza would be most popular with a relative demand of 51%. Arrabbiata is second with 23% preference, while Marinara has 14% and Mushroom has 9% preference. Margherita and Pepperoni have nearly zero demand in this estimate (note: the small sample size makes the estimates somewhat unstable; see Section 10.2.2.5).

With these estimates, we can determine how much of each ingredient to order when we initially stock our restaurant. If we believe these data come from an appropriate sample, we might also replace the Pepperoni offering on the menu because it is estimated to have zero demand; we might consider offering two variations of the popular Arugula pizza. Over time we would gradually update our estimates based on sales data. If we considered adding some new pizzas, we might use MaxDiff again.

10.1.3 Summary of MaxDiff Advantages

In short, MaxDiff does each of the following:

- It assesses the preference for items in an arbitrary list.

- It does this with short, simple tasks on a survey.

- It forces trade-offs, such that respondents can't endorse everything.

- At the same time, it shows equivocal preferences when those exist.

- It goes beyond reporting merely rank order data, and gives magnitude of preference.

- It assesses individual as well as group preference.

Those are strong advantages, but what are the disadvantages of MaxDiff? There are several drawbacks:

1. Respondents must answer multiple trade-off tasks, so the surveys are repetitive even though each task is simple.

2. The surveys must be authored with specialized software that offers the MaxDiff item layout and performs item randomization.

3. Statistical modeling is required to obtain the preference estimates.

4. The results compare items to one another only in terms of relative preference.

5. The requirements for survey authoring and statistical estimation are best met by proprietary, relatively expensive software (Section 10.2.2.6).

Overall, we believe that the advantages of such useful data strongly outweigh the disadvantages for practitioners in industry.

In the next section, we dive into the technical details of how MaxDiff works. After that, we use the `choicetools` package in R to examine a larger data set that simulates users' preferences for activities in a software application.

10.2 Detailed Introduction to MaxDiff Estimation

Now that you have seen an overview of MaxDiff survey items and results, we will review the process in detail. We start by discussing how to write a MaxDiff survey and then look at the statistical model. Later in the chapter we examine how to perform estimation using R code.

10.2.1 Common UX Topics for MaxDiff Surveys

MaxDiff is a good choice when you want to assess user preference among ten or more features, use cases, products, messages, needs, pain points, or other items on a list. Following are some of the things we've seen as topics for MaxDiff surveys. We'll use a hypothetical *videoconference* product to help illustrate some of the topics—not because videoconferencing uniquely fits MaxDiff but only to have a concrete example. Any of the questions on this list might be suitable for a MaxDiff study.

Some of the possible MaxDiff studies are as follows:

- The importance of various features when purchasing a product. Example: the importance of audio and video features for a videoconferencing application.

- Use cases for a product (i.e., things that users want to do). Example: tasks that users would perform with videoconferencing (for work, to see family on holidays, to livestream, etc.).

- Reasons *not* to purchase a product (blocking issues). Example: reasons users might *not* want to have a video call (privacy, bandwidth, etc.).

- Reasons for satisfaction or dissatisfaction with a product. Example: the severity of problems users have experienced with their videoconferencing software.

- Goals for product strategy; what business outcomes matter the most and least to us? (This would be answered by the product team itself, not by users.)

- The priority of engineering efforts for a product; what should we work on? (Again, answered by the product team.)

- Product or advertising messages (what advertisers call "copy"). Example: tag lines for video features such as "full HD" or "clearest video ever."

- Preference for the general product vs. other solutions. Example: preference to communicate via videoconference vs. telephone, email, text, in person meetings, etc.

The following are some other topics that are unrelated to the hypothetical videoconferencing product:

- Amenities or offers for apartments, houses, office work spaces, hotels, automobiles, restaurants, and so forth

- Favorite foods, movies, music, attractions, or vacation destinations

- Activities for employee events

- Places to have meetings or conferences, such as Paris, New York, Tokyo, and so forth

- Employee benefits; benefits for military enlistment

Although MaxDiff would work for any set with two or more items, we suggest a minimum of about ten items before using it. With fewer items—such as preferences among just five features—the complexity of MaxDiff may be too high for both the analyst and respondents. (In those cases, a less complex *conditional logit* model may be appropriate. See Section 10.5 for pointers.)

A crucial point for Quant UXRs with regard to MaxDiff is this: most of the topics listed above occur early in the product lifecycle (see Section 2.2.3). Those stages are often frustrating for Quant UXRs because they are the times when product strategy is determined and yet there may be little or no user data for Quant UXRs to access. MaxDiff offers a way for Quant UXRs to apply their skills in the earliest stages of the product lifecycle, contributing actively to product strategy and direction.

10.2.2 Writing and Fielding a MaxDiff Survey

A MaxDiff survey requires three things:

1. Items that you want to prioritize, as we discuss in the next section

2. A survey platform to field the survey and collect data (Section 10.2.3)

3. Code to estimate respondents' preference (Section 10.2.5)

In the next few sections, we look at how to write the survey. After that, we'll review statistical models and R code for estimation.

10.2.2.1 Writing the Question and Column Headers

Referring to Figure 10-2, the question header is the actual question that respondents answer. In that example, the question is, "If the following 5 pizzas were the only ones available, which pizza would you be MOST likely and LEAST likely to order?"

This question is the most important thing in a MaxDiff exercise because it is the question that *you* want them to answer. It should be clear to them and to you.

A common MaxDiff question header has a structure similar to this:

> Considering only the following [K] items, select the one that is [MOST important] and the one that is [LEAST important] for you when you are selecting a [PRODUCT category].

Or alternatively, similar to the pizza example:

> If you were selecting a [PRODUCT] and the following [K] products were available, which one would you [MOST prefer] and which one would you [LEAST prefer]

The key points in those structures are to clarify for respondents that they should (1) contemplate a hypothetical choice, (2) select a best option and a worst option, and (3) choose those from only the limited set that is shown. Think carefully about it to make sure that it is a question that respondents really can answer (see Section 10.2.2.2).

You will also need to write the *column headers* ("MOST likely" and "LEAST likely" in Figure 10-2). These labels should fit the topic of the survey. You will need an adjective, verb, or adverb ("important," "prefer," "likely," etc.) with paired directional intensifiers ("best/worst," "most/least," etc.).

Feel free to adapt the wording to fit your problem. Here are a few other examples:

- "Most preferred" and "Least preferred"

- "Most important" and "Least important"

- "Best color" and "Worst color"

- "Highest value" and "Lowest value"

10.2.2.2 Developing an Item List

MaxDiff works by taking your list of items, selecting a random few at a time, and forcing users to choose the best and the worst (or most important and least important) from that small random set. Then it repeats the task with new random sets.

This means that you have no control over which items will appear together (or *almost* no control; see Section 10.2.2.3). Your list of items must be suitable for such random selection and respondent choice questions.

The list of items should meet a few criteria:

- You'll want at least ten items or thereabouts (Section 10.2.1).

- The items must be suitable for trade off by respondents. Ask yourself whether it would make sense to make a forced choice between *this item* vs. *that item* for any possible pair of items on the list.

- The items should be short.

- The items should be clear and contain one idea.

Simple Items

A common problem we observe with MaxDiff is items that are too long and complex. This often happens when the survey author mixes different concepts in a single item, such as having both a feature and its benefit, or a conditional situation, or explaining a technical feature.

Consider the videoconferencing example. A team might wish to test the following item: "*Improved brightness in low light conditions so you are always well lit in business meetings.*" That mixes a feature (low light performance), a technical implementation detail (brightness), a benefit (being well lit), and a conditional situation (business meetings).

Because the item is so long, respondents may be confused or read it incorrectly. Also, if they pick it as best or worst, *why*? Is it because of the feature, the implementation, the situation, or not needing the benefit? It is better to present a single idea such as one of these:

- Video is always well lit

- Improved brightness in any situation

- Ideal for business meetings

Those are short, simple, and easy to compare to other items.

Length

Generally speaking, if an item has more than a few words, it is too long. What if you need to explain complex technical scenarios? Explain them separately, not in the MaxDiff exercise, and then use MaxDiff items that are shorthand labels, possibly with mouse-over descriptions if the survey platform will render those correctly for respondents on mobile phones. Because online respondents pay limited attention, a project with long descriptions or technical items may call for fielding in a mixed qualitative-quantitative group interview (Section 10.2.2.7).

Commonality of Items

Another problem arises when the team wants to prioritize items that do not fit a common question. Put differently, sometimes they create a list with items that users cannot answer. For example, the team might want to know whether to target a particular market or to improve performance for all users. This could lead to items such as "improved audio clarity in meetings" (targeting business users) and "improved brightness in any situation" (all users). Respondents may be confused because there is not a clear trade-off. As previously noted, the first item also mixes multiple topics (location and feature) that should be separated; that is often a symptom of items that do not share commonality. Focusing on the exact respondent question (Section 10.2.2.1) will help to avoid this issue.

Developing the Item List

One good way to brainstorm a list of items is to consider different high-level aspects of the experience, and then develop items within each of those categories. For example, with a videoconferencing product, those aspects might be video quality, audio quality, video effects, and so forth. You could first develop several items that relate to video quality, and then do the same for the categories of audio quality and video effects.

Check the Trade-offs

Once you have a list, carefully check that each item makes sense to trade off against the others. It would make no sense to trade off "Ideal for business meetings" vs. "Improved brightness" because brightness might contribute to better meetings. Instead, better comparisons could be "Ideal for business meetings" vs. "Ideal for living rooms," which compares use cases; or "Improved brightness" vs. "Long-range microphone," which compares features.

Maximum Number of Items

How many items can you have? As many as you need. However, when you add more items, you will need to increase the number of tasks (Section 10.2.2.4) or else accept lower precision for each individual respondent (Section 10.2.5.3). It is also difficult to interpret and present results with a large number of items.

Overall, we encourage teams to limit their list to around *35 items* or fewer. When we have seen longer lists—such as a proposed study that had 100 marketing messages— we have found that many of the items could be eliminated through logic ("no, we'd never use that message") or user pre-testing. If you truly must have more items, there are choices about advanced methods that work well with longer lists; see the "MaxDiff Practice" references in Section 10.5.

10.2.2.3 Prohibitions: Items That Can't Appear Together

Occasionally you may have pairs of items that don't make sense to trade off against one another—although each works OK with other items—because one of them is clearly better than the other. In the videoconferencing example, "maximum 720p resolution" and "maximum 1080p resolution" would be an example. Most respondents would always prefer the higher resolution, and when these two items appear together, you are losing potential information from the user because 720p will almost never be chosen as better than 1080p.

In such a case, you may want to use item *prohibitions*, which is an option in some survey platforms (Section 10.2.3). However, we generally recommend either to eliminate such pairs or simply to go ahead and not worry about them (unless you have several or many of them). Any such pair will occur rarely, and respondents can answer such trade-offs even when they are slightly odd. How can you know for sure? Pre-test your survey and see what respondents say.

10.2.2.4 Number of Tasks

Respondents make choices on several successive screens of randomized MaxDiff tasks. But *how many* tasks? The answer depends primarily on the goal of your study: are you looking to get an overall answer for preference across the whole *sample* (and project that to a *population*)? In the earlier pizza example, that is presented in Figure 10-3, which shows average results at the sample level. Or do you want to get very precise estimates for each *individual* respondent, perhaps to use for segmentation or direct targeting? In the pizza example, those are the individual-level results as shown by the circles (and their underlying estimates) in Figure 10-4.

Survey Length for Average, Sample-Level Results

If you are interested primarily in a sample-level estimate, a good practice is to have about six to eight MaxDiff screens, with an *interstitial* screen in the middle to break up the monotony. (An interstitial is a page that tells respondents they're doing well; gives an update on the length of the survey that remains; and, perhaps, shows a cat picture. Most survey software will support the insertion of such pages.) If you have many items—more than about 35—you will want to consider a special sampling method [105].

Survey Length for Individual-Level Results

For precise individual-level estimates, you'll want each item to appear an average of three to five times per respondent on different screens. That way, every respondent will have multiple chances to rate each item as best, worst, or somewhere in the middle, when compared to various other items. That gives enough information for precise individual estimation.

A formula to estimate the number of tasks for good individual-level results is $s = k \times t / m$, where s is the number of screens required, k is the length of your item list, t is the number of times each item should appear across the survey (with a range of 3–5), and m is the number of items on each task. To calculate that, here's an example. The pizza example has 11 pizza varieties. If each one of the 11 items appears 4 times—meaning that each pizza is, on average, compared 4 times to other random sets of pizzas—the survey would need to have 4×11 or 44 total appearances of items. If we show 5 items at a time on each task screen, it would require $44 / 5$ or about 9 task screens to collect the complete set of comparisons.

For simple tasks—such as pizza preference—we might show 6 or 7 items per screen, and thus collect more information with fewer tasks. For instance, if we ask about 7 pizzas at a time, we would need $4 \times 11 / 7$ screens, or about 6 screens, instead of the 9 screens we would need if we show 5 pizza at a time. That would make the survey shorter and less repetitive.

However, that only works if respondents are reasonably able to judge among the larger number of items. For more cognitively complex choices—such as reasons to purchase a product, or preference among technical features—we might show only 3 or 4 items on a screen, and thus require more tasks.

10.2.2.5 Sample Size

How many respondents should you collect? The short answer is this: don't worry too much about sample size. Respondent *quality* and *sample relevance* (reaching the right respondents for your population of interest) are much more important.

As you saw in the pizza example (Section 10.1), MaxDiff works quite well to capture preferences with small samples (in that case, N=6). The question is whether that sample is *representative* of the larger population of interest. The pizza example shows that: do those results (where 10 of the 11 pizzas are vegetarian) represent the population of the United States? Definitely not. But do they represent *Chris's preferences*? Yes! The key is to get the right respondents and have them pay attention to the task.

Beyond that, here are some specific answers:

1. Any number; get whatever data you can

2. N=200

3. N=1000

4. N=100 per group

Let's examine potential reasoning for each of those sample sizes.

- *Any number*: If you can only collect data from a few respondents—such as executive-level decision-makers or any other highly valuable but rare set of respondents—then get them. The data will be better than using opinion alone, assuming that it is fairly representative of whatever group interests you. In this case, consider using an in-person collection method (Section 10.2.2.7). Never let an arbitrary quest for "statistical significance" prevent you from learning from high-quality data.

- *N=200*: The argument for N=200 is that the results of discrete choice surveys such as MaxDiff are usually quite stable for overall averages after surveying about 150–200 respondents [22]. This assumes that you are primarily interested in sample-level results.

- *N=1000*: There are three arguments to collect N=1000. The first argument is that it sounds good and stakeholders are usually satisfied with it. (That is not a great argument, but it reflects our experience.) The second argument is that it will give you enough data to consider latent class analysis or other segmentation efforts (Section 10.5).

If you split a sample of only N=200 responses into several groups, some of the resulting groups likely will be too small for confidence. But with N=1000 the groups should be adequately sized. The third argument is that N=1000 gives enough data to evaluate reliability, such as comparing estimates from split samples.

- *N=100 per group*: This is a variation of the argument for N=1000. If you know in advance that you will split the data into some groups, such as customer or demographic groups of interest, then you will want to collect about N=100 per group. That should yield enough data for reliable estimates for each sample.

Again, we'll emphasize this point: sample *quality* and *representation* are far more important than sample *size*.

10.2.2.6 Survey Fielding Methods

MaxDiff surveys may be fielded in several ways. Most common is a traditional *online panel survey*. First you host the survey online with a fielding platform such as Sawtooth Software, Qualtrics, Conjointly, or others (see Section 10.2.3). Then an email list or third-party sample provider is used to recruit respondents.

Another method is a *computer-assisted telephone interview* (CATI), which can be useful to reach respondents who are difficult or impossible to survey by online methods. In this method, an interviewer calls respondents and records their responses on the survey. For telephone surveys, it is helpful to reduce the number of items per screen to a very small number, because respondents have to remember them and report a choice. We recommend using only two items per task in this case, or possibly three items if they are short and simple.

A *computer-assisted personal interview* (CAPI) may also be used with MaxDiff. This involves a face-to-face interview, such as a field intercept, door-to-door, or in-facility interview. The simplest solution in this case is to hand a device to a respondent and have them complete the MaxDiff questions. Alternatively, it may be conducted similarly to a CATI interview, with the interviewer asking the MaxDiff questions and recording answers.

You might also collect responses using printed, *pencil-and-paper* questionnaires. The problem with this is that MaxDiff involves randomization and you will have to print many different versions, administer them to respondents, and match the randomized designs (the survey version) to each respondent's answers. Platforms such as Sawtooth

Software's Lighthouse Studio make this easier, but it is a time-consuming and potentially error-prone process.

10.2.2.7 Mixed Qualitative-Quantitative Group Interview

One of our favorite methods to field MaxDiff surveys (and conjoint analysis; see Section 10.5) is an in-person, mixed *qualitative-quantitative group interview* (QQGI). The QQGI format combines a qualitative focus group with a quantitative survey performed as part of the group activity.

A QQGI study has several advantages. Because respondents meet you in person, you can research sensitive, proprietary products without worrying that photos or screenshots will be taken by online respondents. If it is important to see, touch, or use a product, you can provide live interaction. Respondents can ask questions and clarify the topic.

You'll observe respondents and know that they pay attention, and thus you will get high-quality data. The respondent mix will be clear, and you can have higher assurance of the representative nature of the group, compared to anonymous online respondents (see Section 10.2.2.5). You can follow up after the survey by asking respondents how they made their choices and add qualitative insights to your analysis of the MaxDiff data. It is also faster to complete than an online survey and you get data immediately.

A typical order of events in QQGI is the following:

1. Introduce the general topic. For example, define videoconferencing.

2. Have respondents discuss their thoughts about the category. For example, they might discuss how they use videoconferencing.

3. Describe some new product ideas, features, or whatever you are researching. Let the respondents ask clarifying questions, but tell them not to share any *opinions* yet.

4. Conduct the MaxDiff interview using tablets, laptops, their own phones, or printed copied.

5. Take a break. During the break, have an associate run the MaxDiff analysis and print out the upper-level results for the moderator.

6. Ask the respondents how they made their choices.

7. Share the high-level results with the group. Ask who liked or disliked the various items at the top and bottom, and probe their answers.

QQGIs provide rich, timely, and specific insights that answer both "how much?" (quant) and "why?" (qualitative) questions.

The primary drawback to QQGI projects is that they are expensive to conduct and involve more time and preparation than posting an online survey. They also require a research team with both qualitative and quantitative skills. Those limitations are offset by the fact that QQGIs yield richer information, very rapidly. The increased costs are partially offset by the fact that they require smaller samples (Section 10.2.2.5).

10.2.3 Survey Authoring Platforms

Although the market changes rapidly and we cannot evaluate all platforms, a few solutions often used for MaxDiff are Sawtooth Software, Qualtrics, and Conjointly.

What is special about those? MaxDiff needs a survey platform that, *at a minimum*, supports both the unique MaxDiff question layout (items with paired "best" and "worst" response columns) and randomization of the items shown as a subset from a longer list. Ideally, a platform will offer balanced randomization, hierarchical Bayes analysis of the responses, individual-level estimation, and possibly other advanced options (see Section 10.5).

Balanced randomization concerns the frequency of appearance of each item, and the two-way co-occurrence of items together. With simple randomization—choosing K items at a time—some items will occur more often than others, and some pairs of items will appear together more often than others due to random chance. Imbalance increases when there are many items, few task screens, or few respondents. This may lead to biased estimates of importance simply because one item appeared more or less often than another, and thus had more chances to be chosen, or appeared more often in combination with another item.

With balanced randomization, the survey platform looks for *design matrices* (the sets of which items were randomized together) where all items appear the same number of times, and appear *together* the same number of times. As of 2022, the best solutions for balanced randomization are offered by Sawtooth Software Lighthouse Studio, including options for evaluating the suitability of the design matrices, generating random test data, and ensuring balanced designs in the presence of item prohibitions (Section 10.2.2.3).

In other platforms, some alternatives to using built-in design optimization are to use a larger sample (imbalance decreases with sample size); to develop the design matrix elsewhere (such as the R flipMaxDiff package [41], or the JMP Discrete Choice Designer [122]) and import that design into your survey platform (Conjointly and Qualtrics offer this, as does Sawtooth Software); or to perform post hoc sensitivity analysis such as repeated split sample assessment of the results. Balanced randomization is especially important when working with smaller samples, such as qualitative-quantitative group interviews (Section 10.2.2.7).

An issue with commercial platforms is that they may be too expensive for some users. John Paul Helveston has developed options to design, author, field, and analyze surveys, including choice-based conjoint (CBC) analysis surveys that are closely related to MaxDiff [58]. At the time of writing, there is not a documented MaxDiff option, although MaxDiff is so similar to CBC that a reader with moderate programming skills could consider a custom adaptation of the code.

Overall, Sawtooth Software currently offers the largest, fastest, and most complete set of tools, options, and analyses for MaxDiff surveys (as well as conjoint analysis) along with support, training classes, publications, and annual academic/practitioner conferences.

10.2.3.1 MaxDiff in Qualtrics

Many UX organizations subscribe to Qualtrics as a default survey platform, and Qualtrics CoreXM offers a MaxDiff item type, as well as a MaxDiff wizard. Our opinion is that— as of the time of writing—these options are suitable for simple or first-time projects. However, if you run many MaxDiff studies, the limitations may be frustrating and you should investigate other options that allow flexibility and additional analyses.

The R choicetools package [23]—which is experimental, unsupported, and in development—offers scripts that enhance the analysis of MaxDiff data from Qualtrics, if a few key points are followed. For the pizza example presented earlier in the chapter, we used choicetools to analyze the pizza data, and we will review choicetools in greater depth in Section 10.3.

In Qualtrics, we have the following recommendations to use the basic MaxDiff item type with `choicetools`. Note that software changes frequently; these are current recommendations as of December 2022.

- Create a single new item and set it to be a "matrix" item with the "MaxDiff" option for layout.

- Import your entire list of items into that MaxDiff matrix, and then use the item settings to display "K at a time" (such as 5 at a time) from the list.

- Set the item to use randomized presentation,

- Set it to have *required* responses (if a respondent can skip a MaxDiff item, the results may be biased because they may choose to skip difficult trade-offs).

- Put the "worst" option as the *first* column, and the "best" option as the *second* column (otherwise, your results will be in reverse order, because the analysis scripts assume that the higher-numbered column is "best").

- Use the preview feature to test the single question several times. Make sure the randomization, format, and required responses are correct.

- Once a single item is working, make multiple copies of it so you have multiple task screens.

- After importing your list of items, do not edit the items later. If you need to change an item, delete all the MaxDiff items and start over (this is due to how Qualtrics exports the data, as of the time of writing in 2023).

- Take the full survey at least once and make sure that your personal choices of "best" and "worst" are recorded properly.

- Then take the survey five to ten times (answering randomly is OK, although even better is to answer realistically and see if the analysis matches your own preferences). Download the data and make sure the analysis scripts work before collecting real respondent data.

- When you export the data, choose the option to save it as a CSV file; select the option to convert the categorical responses to numeric (so "worst" will be coded as 1 while "best" is coded as 2) rather than labels; and be sure to include the randomized design order (the scripts need this to parse the choice data)

- Always pre-test the survey with a few live respondents before fielding it online.

- Use the `choicetools` R code as described in Section 10.3 to estimate and chart your results.

As always, we note that the "think aloud" usability study protocol [100] is very useful for testing surveys before fielding them.

If you have questions about whether exported data are in the correct format, compare their structure to the two example data sets for this chapter (the pizza data and the use case data in Section 10.3). Each of those has numerically coded "best" and "worst" choices along with randomized design matrices in the far rightmost columns. Additional details are provided in Section 10.3.3.

If you use Qualtrics CoreXM for MaxDiff, the most important thing to remember is this: make sure everything is working for a few respondents, all the way through the R analysis to charting the results, before fielding the survey to a larger sample. Do not edit anything in the MaxDiff items between your test and fielding. If you make edits, complete another pilot test before collecting data for your sample.

10.2.4 MaxDiff and Accessibility

When developing any survey, you should pay attention to accessibility and whether the survey is usable by respondents who use screen readers, display magnification, or other accommodations. Because the MaxDiff item type is used less frequently than common item formats such as check boxes, grids, and drop-down lists, MaxDiff items should be reviewed carefully to ensure that they are accessible for respondents.

We have five suggestions to ensure an inclusive respondent experience with MaxDiff surveys:

1. Use a survey platform that supports accessibility options for MaxDiff as well as other survey items.

2. As always, pre-test the survey. Include screen readers and magnification or other options as appropriate in the pre-test.

3. Keep items short, and use smaller sets of items in each task. This helps all respondents but is especially important for those who hold items in short-term memory as they are spoken.

4. If your survey uses images, such as illustrations of features or products, provide appropriate *alt text* (written descriptions that appear or are spoken when the images are not available).

5. Group the "best" and "worst" response columns together, not on opposite sides of the item text. This helps keep them adjacent when they are read, spoken, and magnified.

Additional recommendations, along with notes about accessibility options in common survey platforms, are given in Wainwright and Remy (2022) [145].

10.2.5 MaxDiff Statistical Models

In this section, we describe how the statistical methods work at a conceptual level. For more detailed accounts, see the references in Section 10.5.

10.2.5.1 Counts and Difference Scores

The simplest way to analyze MaxDiff data is simply to find the difference between the proportions of times an item was chosen as "best," minus being chosen as "worst," relative to the number of times it was shown.

The formula is $(B - W) / T$, where B is the number of times an item was chosen as "best," W is the number of times it was chosen as "worst," and T is how many times it appeared. For example, if Arugula pizza appeared 10 times, and was chosen as best in 5 tasks and worst in 1 task, it would score $(5 - 1) / 10$ or 0.4 in relative preference.

There are two primary limitations of a counts analysis. First, it cannot give useful individual-level estimates, because any single respondent will not see any item often enough to have much precision (the denominator T will be very small). The statistical models we consider later achieve higher precision by using transitive information about the relationships *among* items, as well as the overall distribution of preferences across *all* respondents.

The second problem with simple counting analysis is that it doesn't have all of the metric properties you will likely want, such as an interpretable difference score between items or confidence intervals for the estimates. You can do better with a statistical model.

Still, if all you have are item counts, use them. In Section 10.3.4.4 you'll see how to use counts as a quick check before detailed modeling, and how to make a plot of them. Counts are almost always highly consistent with the overall, sample-level estimates obtained from other models. And if you run into difficulty with a complex model, a counts analysis may rescue a study.

10.2.5.2 Multinomial Logit Model

A *multinomial logit model* (MNL) estimates the likelihood of an observed choice that has more than two possible outcomes—in this case, a choice among four or five items shown together. The technical term breaks out as follows: you have *multiple* items that are *nominal* in nature. They are nominal because MaxDiff items are categorical labels, not quantities. Putting those aspects together, the observations are *multinomial*.

What about the *logit* part? The model relates the multinomial values (the items) to the probability that a respondent chose each of them. To find the probabilities, the model finds the coefficients needed for a *logistic equation* that estimates values in a range of 0% to 100%. Any particular probability is determined by a *logistic unit*, for which *logit* is a contraction.

When the statistical model is estimated, it gives a *utility coefficient* (or just "utility" for short) for each item. That is the logit value for the item. The coefficients are related to the probability of choice according to the MNL equation:

$$p(choice) = \frac{e^{util_{oneItem}}}{\sum e^{util_{allItems}}} \tag{10.1}$$

Using that formula, a logit value of 0 corresponds to a probability of 0.5 (50% odds of choice). For values below 0, as they become increasingly negative, the corresponding probabilities become progressively smaller than 0.5, approaching 0 as they grow more negative. On the other hand, positive values indicate increasingly greater than 0.5 probability, approaching 1.0 (100% likelihood of choice) for very large logit values.

Suppose one respondent has a logit coefficient of `util=1.0` for Margherita pizza, while Pepperoni pizza shows `util=-3.0`. You know that Margherita will have a relative choice likelihood greater than 50% because it is a positive value, while Pepperoni will be below 50% because it is negative. You can use the MNL formula to estimate the exact likelihood that respondent will choose a Margherita pizza as opposed to a Pepperoni pizza.

In R syntax, the MNL probability estimate is `exp(1.0) / (exp(1.0) + exp(-3.0))`, which is a value of 0.982. This respondent would have a 98% likelihood of choosing a Margherita pizza rather than a Pepperoni pizza (which would be 2%). To see the likelihood when choosing among a larger set of items, add the coefficients for each of the items to the denominator of the equation. There is an example in Section 10.1.2 earlier in the chapter.

An MNL model estimates the coefficient values for every item in your sample as a whole. If you use a *multilevel* (aka *mixed effects* or *hierarchical*) model, it will also find the estimates for every respondent, as discussed in the next section.

The key points are these: the MNL model gives a mathematical estimate for each item, with an exact definition and metric properties. It relates that item to the observed likelihood of choice in your data. For much more information on these models, see Sections 9.3, 9.4, and 13.5 in the R companion book [25], and this chapter's "Learning More" section (Section 10.5).

10.2.5.3 Hierarchical Bayes Model

A *hierarchical Bayes* (HB) model [120] is the most popular way to find the estimates for the MNL model of choice probabilities. The *hierarchical* part of the name means that it gives estimates for the sample as a whole (the "upper-level model") and for every respondent (the "lower-level model"). They are "hierarchical" because the sample is hierarchically composed of the respondents.

The *Bayes* part of the model refers to Bayesian estimation procedures, named after an early statistician. Bayesian estimation finds the most likely values for items' coefficients through an iterative process. We won't say much about HB models here, except for one important part: they can be slow to run, and may take minutes, or—in the case of large data sets—hours to run.

We estimate MaxDiff values using HB models, which are the default in Sawtooth Software and many other survey platforms. The estimated values can be used in the MNL formula (presented in the previous section), if desired.

If you're curious to learn more about Bayesian models, see the references in Section 10.5.

10.2.5.4 Using and Reporting the Scores

The statistical model will give you estimates of the *utility* values (MNL coefficients) for every item. Many platforms give you *raw* utility values that are the exact values to use in the MNL equation (Section 10.2.5.2).

However, it is relatively rare in practice to use the MNL equation to compare the preference share of MaxDiff items! It is far more common to use and interpret *charts* of the values, as we did early in this chapter to compare pizza preferences (see Figures 10-3 and 10-4). Charts are much easier to interpret and explain.

There is one problem with those charts: the raw (MNL coefficient) estimates are typically modest in magnitude, often less than ±2.0. It may be difficult to impress stakeholders with a difference of 2.5 vs. 1.5 on a chart...although it reflects a 73:27% difference in choice likelihood (per the MNL formula, in R: exp(2.5) / (exp(2.5) + exp(1.5))). It is also difficult to explain what those values *mean* because stakeholders are rarely interested in the MNL formula.

To make charts more interpretable, analysts often transform the utility values. Common options are to use *zero-centered difference* scores, which are usually much larger in magnitude, such as 6.3 or 14.5, or *probability scale* estimates that give the odds of each item being chosen vs. the average of all items.

Rather than worrying too much about the details of the scores being reported, our recommendation is this: use the default scores that your platform or code reports, and figure out how you will explain them to your stakeholders. You might multiply values by a constant such as 10 or 100 if that makes the chart clearer.

We suggest that you *avoid* explaining the estimates technically because that will be a distraction with most audiences. What we usually say is something like this: "These are relative scores for the importance of each item, compared to the others. Higher scores mean an item was more preferred, while lower scores mean it was less preferred. The underlying models are complex to explain, but the good news is that we don't need to get into that math to see how strongly the items were preferred."

To learn more about options for rescaling MaxDiff coefficients, see the reference materials from Sawtooth Software [126].

10.3 Example: Information Seeking Use Cases

Now that you understand the concepts of MaxDiff, we'll walk through estimation in R using the `choicetools` package and a data set in the Qualtrics CoreXM format.

The example in this section reflects a common UX application of MaxDiff: assessing the comparative importance of end-user *use cases*—tasks that users want to accomplish with a product. In this case, we consider the importance of general areas of *information content* that users might seek online in a search engine such as Google or Bing, or in a dedicated application such as a banking app. As in other chapters, these data are structurally identical to a real data set but contain simulated responses.

The following sections are complex, and this is a good time to remind you that this level of code and estimation is only needed if two things are simultaneously true: (1) you want to run a MaxDiff study and obtain HB estimates and charts similar to those in this chapter; and (2) you are using a platform such as Qualtrics that does not include those.

Platforms vary, and some do not require following any of these steps in R. In particular, Sawtooth Software Lighthouse Studio includes a complete set of options for estimation and does not require using R. Many analysts plot its data in Excel (although you could always import it into R for more complex analysis or charting).

On the other hand, Quant UXRs often like to know exactly what's happening with their data and analyses. The following sections have all of that detail.

10.3.1 Overview: MaxDiff for Information Seeking

Suppose you want to answer this question: when users go online, which activities are most important to them? You might want this information if, for example, you are designing a laptop computer, an operating system, or a search engine. It might also be important to understand the degree of interest in one category of information, if you are creating an application for it.

One way to approach such a question is through user log data—such as search queries they issue or how much they use various applications (see Chapter 9, "Log Sequence Visualization"). But there are several limitations to logs analysis:

- Logs tell you what users are doing but not what they *want* to do. If there is something users would like to do, but it is blocked or made difficult by the UI or technical design, it does not show up in the logs.

- Logs are granular and difficult to *aggregate* into higher-level themes, such as general categories of behavior, needs, or information. It is often simpler to ask users to report preferences or needs at those higher levels.

- You only get log data *after* making a product. If you want to know what to do in advance of making the product, you need to inquire in other ways.

- Logs data is only available for *your* products. If you are interested in user behavior, wishes, or preferences with regard to competitors' products, you will not have logs. Also, if you are creating a platform such as hardware or an operating system, you usually will not be able to access logs for other products that run on top of your platform (such as applications or plug ins).

MaxDiff makes it possible to complement logs data with higher-level insights and to understand user needs in advance of creating a product. This expands the range of analyses Quant UXRs may perform across the lifecycle.

10.3.2 Survey Format

Our example MaxDiff survey asks the following question:

> Consider the information, content, and activities you use online.
> Among the following 5 areas, which one is MOST important to
> you, and which one is LEAST important?

As we described in Section 10.2.2.1, the question header should be directly related to the research question. In this case, our research question is very general, wanting to assess users' perceived importance of information types, regardless of how they access the information or on what device. We could be interested in this general level of importance if, for example, we wanted to understand *users* in themselves, apart from any specific platform.

On the other hand, if we wanted to answer questions about what users are actually *doing* with a specific *platform*, we could ask a different question, such as:

> Consider the information you have accessed in the past week
> using a smart speaker (such as Google Home or Amazon Echo).
> Among the following 5 areas, which one have you accessed MOST
> OFTEN and which one have you accessed LEAST OFTEN (if
> at all)?

If we wanted to know about *frustration* with accessing information on a *smartphone*, we might ask something like the following:

> When you are looking for information on your smartphone, which
> of the following areas is the EASIEST to find on your phone, and
> which is MOST DIFFICULT to find?

Also, we would change the response labels to match the question about "important," "often," "difficult," and so forth.

The key point is that the question and items should be closely tied to a specific research question. Assuming that has been done, and that we are interested in general perceived importance of use cases, we can now look at the details of the survey.

The following 19 items are in the MaxDiff response set:

Banking	Local events	Religious information	Sports news
Casual gaming	Multiplayer gaming	Restaurant reviews	Travel reservations
Driving directions	Music	Schoolwork	Video meetings
Email	Political news	Shopping	Weather forecast
Entertaining videos	Recipes	Social media	

This list is only an example; the items are not necessarily complete or worded appropriately for a real project.

The final MaxDiff survey was defined with the following options:

- K=19 items (per preceding table)

- T=4 appearances of each item, for each respondent

- M=5 items shown on each task screen

- S=15 total screens, which was calculated as 19 items × 4 appearances / 5 item per screen = 15 screens (see Section 10.2.2.4)

An example task format is shown in Figure 10-5. For a real survey, you might want to use fewer than 15 screens, or put interstitial screens between sets of a few tasks each (Section 10.2.2.4).

Consider the **information, content, and activities you use online**. Among the following 5 areas which one is MOST important to you, and which one is LEAST important?

1 / 15

LEAST important		MOST important
○	Multiplayer gaming	○
○	Restaurant reviews	○
○	Banking	○
○	Entertaining videos	○
○	Weather forecast	○

Figure 10-5. *An example task for a MaxDiff survey about online information access*

The simulated response data set is `qualtrics-maxdiff-usecases.csv`. Later in this chapter we'll show how to load the data into R to estimate users' preferences.

10.3.3 Data Format

The simulated data are in a Qualtrics CoreXM export format, as of the time of writing. Details on setting up MaxDiff in Qualtrics, and exporting the data, are in Section 10.2.3.1.

The `choicetools` package expects data to come in a specific "wide" format with one row of observations per respondent and repeated columns for every item on every task screen. There are six important aspects to that format:

1. The data should be *complete*, with every task answered by every respondent. Either set the data export to "complete" respondents or filter them later. This is important both for the scripts and for the underlying method; if respondents are incomplete, they may have discontinued due to difficult trade-offs, and you want to avoid that as a source of potential bias.

2. There must be respondent *identifiers* in the first column. These are used to index the respondents during estimation.

3. The data should be *sparse*, with only two items that have responses for each task, and all other items' columns blank. These two are the "best" and "worst" items for each task. Figure 10-6 shows an example of the sparse data.

4. The best item should be coded as "2" and the worst item coded as "1". That is so the estimation will know which direction corresponds to "better," namely as the one with the higher value.

5. There must be *design matrices* in the file that show all of the items that were considered (including, and additional to, the two selected as best and worst). Figure 10-7 shows an example of design matrices.

6. There are three important *header rows* of information that include the item number, the text as shown to respondents, and internal coding information. The `choicetools` scripts expect those three rows exactly as output by Qualtrics CoreXM.

	A	B	C	D	E	F	G	H	I
1	ResponseID	Q1_1	Q1_2	Q1_3	Q1_4	Q1_5	Q1_6	Q1_7	Q1_8
2	ResponseID	Consider the⯈	Consider the⯈	Consider the⯈	Consider the⯈	Consider the⯈	Consider the⯈	Consider the⯈	Consider the⯈
3	{'ImportId': 'r⯈	{'ImportId': 'Q⯈	{'ImportId': 'Q⯈	{'ImportId': 'Q⯈	{'ImportId': 'Q⯈	{'ImportId': 'Q⯈	{'ImportId': 'Q⯈	{'ImportId': 'Q⯈	{'ImportId': 'Q⯈
4	1						2		1
5	2	2							1
6	3		2						
7	4	1							
8	5	1							

Figure 10-6. *Sparse data in the Qualtrics export format for MaxDiff tasks. Items chosen as best are coded with 2 and those chosen as worst are coded as 1, while all other items are blank. For example, the first respondent chose item 6 as best and item 8 as worst*

If you are using Qualtrics CoreXM, it is crucial to check the data completeness, format, and coding. For other data sources, see the next section (Section 10.3.3.1).

KA	KB	KC	KD	KE
DO-Q-Q1	DO-Q-Q2	DO-Q-Q3	DO-Q-Q4	DO-Q-Q5
Display Orde▸	Display Orde▸	Display Orde▸	Display Orde▸	Display Orde▸
{'ImportId': ' ▸	{'ImportId': ' ▸	{'ImportId': ' ▸	{'ImportId': ' ▸	{'ImportId': ' ▸
4|6|8|14|16	3|8|11|12|13	3|4|15|17|19	3|5|10|16|18	3|6|7|9|14
1|4|8|11|18	5|8|14|15|16	1|2|9|10|14	4|7|12|14|19	3|11|13|14|1▸
2|4|9|17|18	5|9|11|14|19	1|3|5|13|17	5|6|10|15|18	2|4|5|7|16
2|6|8|13|14	1|5|6|12|19	2|9|12|15|16	8|11|12|17|1▸	3|4|12|13|14
2|8|10|17|18	5|7|15|18|19	2|6|7|13|14	3|7|10|11|16	4|7|8|12|17

Figure 10-7. *Design matrices in the Qualtrics MaxDiff export. Each design order column shows the item numbers for the 5 items displayed on a particular task. For example, for the first respondent, the first task showed items 4, 6, 8, 14, and 16, drawn randomly from the set of 19 items (not necessarily in that order)*

The choicetools scripts perform some basic checks of data quality, but it is best to check your data format carefully before using them. In the exported data files, many elements—such as the item names—are not clearly identified and the scripts must infer them. Various issues with data sets may confuse the import scripts. Clean data will give you a head start on debugging any issues that arise.

10.3.3.1 Data Sets in Other Formats

What if you use another platform such as Sawtooth Software or Conjointly?

First, you may not need to use the choicetools package. Sawtooth Software's Lighthouse Studio suite includes highly optimized hierarchical Bayes estimation. We recommend using that if you field a study with Sawtooth Software. You could then pick up with our R scripts *after* the estimation steps if you wish to use the choicetools plotting functions.

Another alternative is to export the data and write code to translate your data into the format expected by choicetools. In the case of Sawtooth Software, you could export a MaxDiff section in the "CHO file" format; see the in-product help files.

If you wish to translate data from another format to use with choicetools, the example data sets qualtrics-maxdiff-usecases.csv and qualtrics-pizza-maxdiff.csv demonstrate the structural elements that are needed to mimic the layout, response coding, design matrices, and column headers. In this chapter's R code file, there is additional code (at the end of the file) to create Qualtrics-style headers for arbitrary data, and to create simulated data sets. You could adapt those scripts to your MaxDiff structure and then replace the data with your data. Section 10.3.4.3 demonstrates how to load the data for choicetools.

10.3.4 Estimation with the **choicetools** Package

Once you have data in the expected format as exported by Qualtrics CoreXM (section 10.2.3.1), you are ready to estimate and examine the results using the choicetools package.

10.3.4.1 Setup for Estimation

The first step for many first-time analyses in R is to install necessary packages. This is often an iterative process where you discover additional packages needed as you work through code. For a start, install the following packages:

```
install.packages(c("reshape2", "ggplot2", "mlogit",
  "ChoiceModelR", "Rmisc", "matrixStats", "superheat", "corrplot",
  "ggridges", "devtools"))
```

Those packages may install other packages that they need. These packages are used to convert data formats (reshape2 [56]), estimate statistical models (mlogit [35] and ChoiceModelR [129]), plot results (ggplot2 [148], superheat [3], corrplot [147], and ggridges [152]), install packages (devtools [150]), and perform other mathematical functions (Rmisc [61] and matrixStats [5]).

Watch the R console during the installation. If you are asked to install other packages, choose "yes." However, if you are asked whether to "compile a package from source," it is usually best to choose "no."

Next, install the choicetools package from GitHub:

```
devtools::install_github("cnchapman/choicetools")
```

If you get an error, you may need additional components for your R setup, depending on the version and your operating system. To check whether choicetools is working, try this command:

```
packageVersion("choicetools")
```

If the library was installed, R will display its version (which will be 0.0.0.9083 or, probably, higher).

This is a good time to remind you that `choicetools` is experimental and unsupported. For alternatives, see Section 10.5.

10.3.4.2 Check the Data

You begin by loading the `choicetools` package and downloading the use cases data file (if needed):

```
library(choicetools)
if (!file.exists("qualtrics-maxdiff-usecases.csv")) {
  download.file(url = "https://bit.ly/3SRnq9l",
                destfile = "qualtrics-maxdiff-usecases.csv",
                method="auto")
}
```

If you don't have a live Internet connection, you need to download the data file separately from the book's website at `https://quantuxbook.com`.

As previously mentioned, the `choicetools` scripts must infer everything about your MaxDiff study from the structure of a Qualtrics CSV file. The `parse.md.qualtrics()` function does this. Run it and read its output carefully, as in the next command. Be sure to add a folder path for the data file if you downloaded the data separately from using the `download.file()` command above.

```
# change the file location to match your system as needed
parse.md.qualtrics("qualtrics-maxdiff-usecases.csv")
```

It is important to read all of the output from the parsing function and make sure it matches your study design. We'll step through each portion of the output.

First, make sure it identified the question and all of the items:

```
Found MaxDiff question header:
"Consider the information, content, and activities you use online.
Among the following 5 areas which one is MOST important to you, and
which one is LEAST important?"

Qualtrics 'legacy' file format detected. Parsing.
File structure implies 19 MaxDiff items.
```

```
Found K = 19 MaxDiff items (column text after the header above)
 "Banking"              "Casual gaming"        "Driving directions"
 "Email"                "Entertaining videos"  "Local events"
  ...
```

That matches the survey design and correctly lists the 19 items. Problems may arise when there are multiple questions in the data file with randomized response options. In those cases, remove those items' design order columns.

Next, did it find all the tasks and corresponding design matrices?

```
Found M = 15 screens of MaxDiff items per respondent.
Columns with experimental design matrices are:
 [1] 287 288 289 290 291 292 293 294 295 296 297 298 299 300 301
```

This is correct, finding 15 tasks and design matrix columns. In this case, the design order columns are in consecutive order, but that may vary in other data files.

Did it find the correct number of respondents?

```
Observations are in rows 4 to 203. Found N = 200 respondents.
Checking design matrices ...
Found 5 (double-check: OK) items shown per task.

Reviewing coded answers.
Found min code (worst?) = 1 and max (best?) =  2
Found average = 15 'best' answers, and average = 15 'worst' answers.
```

It found 200 complete respondents, and confirms that 5 items were shown on each task. The coding for the best and worst directions looks appropriate, and every respondent answered 15 tasks.

Finally, it gives a summary at the end:

```
Found N = 200 complete responses, and N = 0 with missing observations.
 ...
==> Your data appear OK in this check.
```

If there were incomplete respondents, it would be best to investigate and remove them if necessary. As is, the next step is to load the data.

10.3.4.3 Load the Data

The choicetools library is designed to use a single list object to hold all aspects of a MaxDiff study. We'll call this object md.design to agree with the package's examples, although it could have any name.

You can think of this list object as being like a filing cabinet. First we'll put in the study design, then the data, and then the results. By putting them all in one object, it is easy for functions to find what they need to perform estimation or make charts.

Comments in the choicetools code files describe how to create the md.define object, but that level of detailed specification is not necessary in most cases. If your data parse correctly as shown in the previous section, then the function parse. md.qualtrics() will set up the study object for you. Just add the argument return = TRUE and assign the result to a new md.design object. As always, change the folder location if needed. The code is

```
md.define <- parse.md.qualtrics("qualtrics-maxdiff-usecases.csv",
                                returnList = TRUE)
```

The function repeats information we saw previously, along with suggested R code to estimate the model using the md.quicklogit() function:

```
Reading file: qualtrics-maxdiff-usecases.csv
  ...
Example code snippet to use it:
  ...
  mod.logit           <- md.quicklogit(md.define)
```

If you inspect md.define you can see how it set up the study design. Here are some excerpts from that object:

```
> md.define
$file.qsv
[1] "qualtrics-maxdiff-usecases.csv"
$md.item.k
[1] 19
$md.item.tasks
[1] 15
$md.item.pertask
```

```
[1] 5
...
$md.item.names
 [1] "Banking"            "Casual gaming"
 [3] "Driving directions" "Email"
...
```

It knows where to find the data file, how many items are in the study, the number of tasks, the item labels, and so forth. These are used to read and structure the data in the next step.

We use the study definition to load the data and put it into an element called `md.block` within the `md.define` object:

```
md.define$md.block <- read.md.qualtrics(md.define)$md.block
```

```
## Found MaxDiff question header:
## "Consider the information, content, and activities you use online. Among
the follow
##
## Qualtrics 'legacy' file format detected. Parsing.
## Found average = 15 'best' answers, and average = 15 'worst' answers.
## Of N = 200 total:
## Found N = 200 complete responses, and N = 0 with missing observations.
## Recoding 200 complete responses. (Dropping 0 incomplete.) ... 100
200 done.
```

The output confirms that N=200 respondents were found, with 15 tasks each.

The data are in a data frame called `md.define$md.block`.

If you inspect `md.define$md.block` (omitted here), you'll find that it is in *long* format. This is in preparation for using the `ChoiceModelR` package [129] for estimation (Section 10.3.4.4). In long format, it has ten rows for each task—five rows that code a tasks' items for the *best* direction, plus five rows for the items in the *worst* direction—along with the task ("block") number, the respondent identifier, and the items that were chosen as best or worst in the `win` and `choice.coded` columns. (If your data come from another source, you could also set them up to match this format and then use `choicetools` for estimation.)

10.3.4.4 Estimate the Model

Hierarchical Bayes (HB) estimation can take a while—from minutes to hours, depending on the data size—and it's helpful to get a quick check first. This lets you see whether estimation works at all and whether your items' best and worst directions look appropriate.

A fast way to do that is to plot the item *counts*—how many times each item was chosen as best or as worst—along with the net difference between them (see Section 10.2.5.1). The function plot.md.counts() does that, as shown in Figure 10-8:

```
plot.md.counts(md.define)
```

The items you expect to be best should be at the top, and the items that are worse should be near the bottom. In Figure 10-8 shows that information use cases related to social media, entertainment, music, and sports were the most popular.

If the items appear to be in reverse order, check the numeric values in your data set for the *best* and *worst* codes. The code for *best* choices should be larger than that for *worst* choices; if it is smaller, replace it with a larger value.

If the counts look good, we are ready for full HB estimation. A crucial part about HB estimation is to set an appropriate number of iterations—how many times it works to incrementally improve a model. For most MaxDiff projects, a good target is 10000 or 20000 iterations. We'll say more about that in Section 10.3.4.5.1, but for now, let's use 10000 iterations. The process also involves an element of random sampling across the iterations. To make that consistent, we set a starting point when calling the estimation function (using the mcmc.seed argument in the following code).

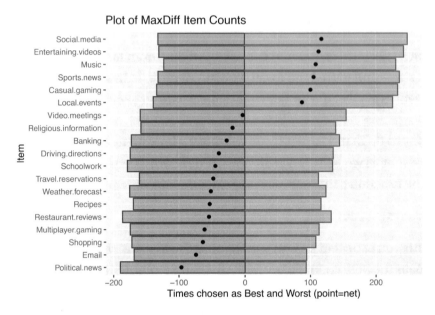

Figure 10-8. *A plot of the item counts as the number of times chosen as best or worst, and the net difference between those. This is useful as a quick check on the data before hierarchical Bayes estimation*

In this code, we estimate the model and put it into a separate object `test.hb`. Then we copy the estimated utilities from there into our `md.define` study object. The HB process may take a few minutes, depending on your computer:

```
test.hb <- md.hb(md.define, mcmc.iters = 10000,
            mcmc.seed = 98101)                    # estimate
md.define$md.model.hb    <- test.hb$md.model      # save results
md.define$md.hb.betas.zc <- test.hb$md.hb.betas.zc # individual scores
```

10.3.4.5 Plot the Results

What did we find? The function `plot.md.range()` shows the sample (upper level) averages with 95% Bayesian credible intervals (the central 95% range of estimates, a Bayesian counterpart to traditional confidence intervals). We plot those and add an informative axis label:

```
plot.md.range(md.define) +                    # plot upper level averages
  theme_minimal() +
  xlab("Information Use Cases")
```

The result is shown in Figure 10-9. The top six items—social media, entertaining videos, music, and so forth—are roughly tied for preference, followed by a gap and then video meetings. Then there is another gap, and the remaining items show a long tail of decreasing interest. As we saw with the pizza data at the start of the chapter, this is a relatively common pattern in MaxDiff data—to have a few strongly preferred items, perhaps in a couple of separate groups of items, and then a long tail of items in the middle or lower area of preference.

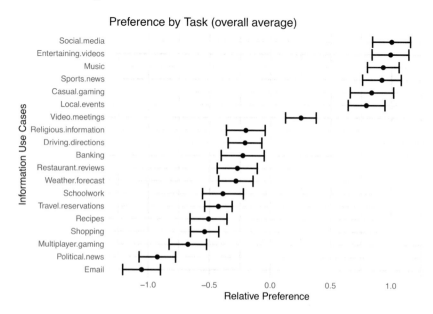

Figure 10-9. *Average preferences for the simulated information use case data, N=200*

There is a side note for those fluent in R: the `choicetools` plot functions return ggplot2 objects. You can assign those to objects and add `ggplot2` elements or themes as you wish. In the code snippets here, we change the theme to `theme_minimal` and add axis labels; more complex manipulation is also possible. (You can also access the exact R code for the functions in `choicetools`.)

As always, we check the individual-level distribution using `plot.md.indiv()`:

```
plot.md.indiv(md.define) +                    # individual distributions
  theme_minimal() +
  ylab("Information Use Cases")
```

The individual distributions are plotted in Figure 10-10. At the individual level, the distributions are quite similar for the top three items. Looking farther down, there is a somewhat stronger upper tail of respondents interested in casual gaming. After that, the lower items have relatively few respondents who show especially strong interest (a right-hand tail on the distribution), but on the other hand, also relatively few who are especially disinterested.

How would we use the individual-level distributions? One way—which one of Chris's PM colleagues does—is to use the individual scores to identify respondents who are especially interested or disinterested in a particular use case and contact them for interviews. If users are highly interested, they are good candidates to give feedback on new designs that address their specific needs. Users who are disinterested may reveal issues that the team has not considered. More generally, we can use the distributions to answer questions about how many respondents have especially high (or low) interest.

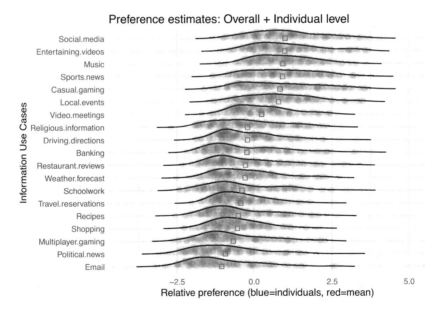

Figure 10-10. *Distribution of estimated individual preferences for the information use case data, N=200*

More on HB Iterations

The HB method requires setting a specific number of iterations to run the estimation procedure. How do we determine that? This is a complex issue in HB discussion (see Section 10.5), but we offer two general recommendations.

The first recommendation is to inspect the *trace plot* that is shown while the HB estimation runs (which is similar in both `choicetools` and Sawtooth Software Lighthouse Studio). This plots the running averages of the estimates for each item's coefficient, with a separate colored line for each item. If it runs for 10000 iterations or longer *and* the plot is relatively flat in the latter 25% of the traces, then we may feel confident that the estimates are good to use. In HB jargon, they have *converged*. Figure 10-11 shows an example of a well-converged trace plot.

For comparison, Figure 10-12 shows a *non*-converged traced plot, where the estimates are not flat, but are blowing up and changing.

In such a case, you should run another 10000 or 20000 iterations. You can instruct `choicetools` to add iterations rather than beginning from scratch by adding the parameter `restart=TRUE` when calling `md.hb()`.

Our second recommendation is this: if you run the HB process for 50000 or more iterations, and it still has not converged, then you need to decide whether the data are reasonable enough to use. In general, we believe that the value of data is so high that it should be used for decisions even when it is imperfect.

However, that assumes that you believe the data represent realistic user preferences, gathered from an appropriate survey. If the respondents were of low quality, sped through the survey, answered long and complex items, or answered about things they don't know or don't care about, then the data should not be used. In fact, in any of those cases, the estimates shouldn't be used whether the trace plot converges or not, although the presence of non-convergence may be a useful signal to reconsider the quality.

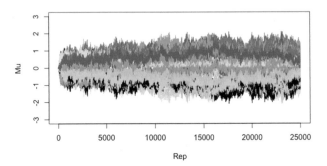

Figure 10-11. *An HB trace plot that is well converged in the latter quarter of the plot, from iterations 16000–25000. The trace plot expands in range until about 13000 iterations, and then levels off. After that, it shows a consistent range with individual lines varying around horizontally flat values. These estimates are suitable for reporting and interpretation*

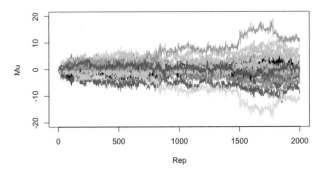

Figure 10-12. *An HB trace plot that has not yet converged. Starting at 1500 iterations, the scale increases and the estimation lines for individual items are far from being flat and centered on consistent values. For a trace plot of this kind, run another 10000 or 20000 HB iterations*

For more on HB convergence, including more formal tests of convergence, see the HB references in Section 10.5.

10.3.5 Next Steps

After running a MaxDiff study, the following are usual next steps:

1. Interpret the results for the stakeholders. We don't recommend sharing raw results because the process and utilities are not interpretable without explanation or training.

2. Follow up with qualitative research to answer any questions that remain about preference. If there is a surprise in the results, qualitative interviews can clarify it.

3. Consider latent class analysis to investigate whether there are meaningful groups who have patterns that are similar to one another while being different from other groups (see Section 10.5).

10.4 Key Points

This has been a long chapter covering a unique survey method, specialized statistical models, and experimental R code. The key points are as follows:

- When you want to prioritize a list of features, benefits, messages, blocking issues, or use cases , MaxDiff is worth consideration (Section 10.2.1).

- MaxDiff is a valuable tool for Quant UXRs who are working early in the development process. It can yield interesting insight, using modestly advanced statistical methods, for many problems that cannot be tackled by logs analysis or other methods that depend on data from a functional product (Section 10.3.1).

- MaxDiff is preferable to typical Likert scale ratings and stack ranking methods because it gives estimates with stronger metric properties (Section 10.1.3).

- Key requirements include a list of roughly 10 or more items that are suitable for trade off vs. one another; a survey platform that supports MaxDiff items; and estimation methods given by your survey platform or by using custom code in R or another language.

- Hierarchical Bayes estimation gives results for both the sample as a whole (the upper-level model) and for each respondent (the lower-level model). This is helpful to investigate the distribution of responses, and potentially may be used for segmentation or targeting (Section 10.3.4.5).

- Respondent quality is much more important than sample size. If you know that the respondents are representative, and they pay attention to the tasks, MaxDiff works with modest samples (Section 10.2.2.5). Online sample requirements start at about N=200, while in-person samples may be much smaller.

- Always pre-test the items for clarity and respondent understanding, and, if you are using custom code, to ensure that the code works (Section 10.2.3.1).

10.5 Learning More

This chapter has covered a large range of materials and concepts, including practical advice on constructing MaxDiff surveys, discussion of the underlying statistical models, and R code to analyze MaxDiff data.

Each of those areas has extensive literature from which you can learn more. Unfortunately, the literature on choice models is too often excessively and unnecessarily mathematical. The following pointers will help you get started at a level that is appropriate for your interests.

MaxDiff practice. A friendly introduction to MaxDiff for UX research applications is Luster [87]. The single best reference for practitioners using MaxDiff is a book by Keith Chrzan and Bryan Orme, *Applied MaxDiff* [33]. It focuses on practical implementation and analysis of MaxDiff studies in industry settings. Beyond that, the Sawtooth Software Conference—now known as the Analytics & Insights Summit—publishes formal proceedings volumes each year. There are usually several papers with interesting guidance, extension, and discussion of MaxDiff best practices. The *Proceedings* are available for free online [125]. The annual conference itself includes hands-on training and showcases practitioner and academic advances in choice modeling generally. A technical reference for MaxDiff is *Best-Worst Scaling* by Jordan Louviere and colleagues [84]. Louviere invented MaxDiff.

Multinomial logit models. There is abundant literature on multinomial logit (MNL) models. One starting place is Section 13.3 in the R companion book [25]. Although that section focuses on conjoint analysis (see below), the models are mathematically identical. For technical coverage, see Chapter 11 in Hilbe's *Logistic Regression Models* [60]. The HB references discussed in the next paragraph also cover MNL models.

Hierarchical Bayes models. There are two separate aspects to hierarchical Bayes (HB) models: the hierarchical aspect known as a mixed-effects model, and the Bayesian estimation aspect. Both of those are introduced in the R companion book [25] in Sections 9.3 (hierarchical models), 9.4 (Bayesian hierarchical models), and 13.5 (HB models for choice data). Kenneth Train's book, *Discrete Choice Methods with Simulation* [140], gives especially readable (although technical) coverage of all aspects of general discrete choice models, of which MaxDiff is a derivative. This includes HB and other methods of estimation.

Segmentation and latent class analysis. MNL and HB estimation generally assume that the respondent population is normally distributed (in a multivariate sense). Some respondents will have high or low interest in any given item but their preferences are mapped to a common distribution across respondents. A question you should consider is the following: what if they are not part of a common distribution but come from *separate* groups that have systematically different sets of preferences?

Analysts often explore that question by applying clustering methods to the individual-level estimates from MaxDiff. This may be done with any generic clustering method (see Section 11.3 in [25] or Section 10.3 in [127]). However, individual estimates are noisy, and a preferable technique is *latent class analysis* (LCA). LCA investigates whether there are systematic patterns in the choice responses themselves, and estimates the item utility values simultaneously with identifying clusters.

For a general introduction to LCA with choice data (applicable to conjoint analysis and MaxDiff), see Ramaswamy and Cohen [110]. For users of Sawtooth Software, see the *Latent Class Manual* [104]. An R package for latent class analysis of multinomial choice data is the `gmnl` package [121].

Conjoint analysis. Choice-based conjoint analysis is a popular technique to determine product pricing, feature preference, market demand, and similar questions for products in a competitive market. Whereas MaxDiff works with a list of features or the like, conjoint analysis works by asking respondents to choose among *products* that combine multiple randomized attributes at a time, such as a brand, several features, and a price. The underlying logic, math, concepts, and survey methods are quite similar (and in some cases identical) for conjoint analysis and MaxDiff. If you find MaxDiff useful, you'll likely enjoy conjoint analysis as well. Two starting places are Chapters 9 and 13 in the R companion book [25], and the excellent, practitioner-friendly introduction by Bryan Orme, *Getting Started with Conjoint Analysis* [106].

10.6 Exercises

The following exercises will take you through the process of developing and fielding a MaxDiff survey. Exercises 1, 2, and 3 are similar to activities in classes that Chris taught internally at Google. Exercises 4 and 5 are for readers who are proficient in R. They can help you understand more about how the models work and what to expect from noisy data such as inattentive responses from survey participants.

1. Write your own MaxDiff exercise for something of interest to you. Write the exact question header and brainstorm a list of items to trade off. Implement it in a survey tool (Sawtooth Software and Conjointly each offer free trials). Here are some ideas: food preferences; apartment amenities; benefits at work; places to go on holiday; favorite movies; activities for a fun event at work; best US president; and social issues. (Be careful asking anyone about politics, of course!)

2. Test your survey with a few friends or family using think-aloud protocol [100]. What would you want to change before actually fielding your survey?

3. Run the survey for N=10 or more respondents. Estimate the utilities using either a method supplied in the survey platform or described in this chapter. What do the results tell you? Are there any surprises?

4. *Moderate code.* Adapt the code in the appendix of the R file for this chapter and simulate MaxDiff data for a different problem. Change the values used to simulate the preference coefficients and generate new data. Estimate the results. How well do the results match the expected values you set?

5. *Advanced code.* Add random responses to the simulation data for Exercise 4 and re-estimate the model. How well does it perform when adding 1N random responses (i.e., if you have N=200, add another N=200 with random answers)? How about 5N random responses (i.e., adding N=1000)? 10N? At what point is the MaxDiff signal obscured so much by the added, random responses that you are unable to recover the values that were simulated?

PART IV

Organizations and Careers

Introduction to Part IV

The chapters in Part IV examine organizational issues for Quant UX research. In Chapter 11, "UX Organizations," we review common patterns for structuring research teams with Quant UX researchers. We recommend that UX managers, in particular, pay attention to the advantages and common problems of the various models. This chapter will also help industry newcomers, as well as readers who are changing roles or companies, to understand various organizational structures.

Chapter 12, "Interviews and Job Postings," describes how Quant UXRs are commonly hired. We share advice that we hope will improve interview experiences for both candidates and interviewers.

This is naturally followed by Chapter 13, "Research Processes, Reporting, and Stakeholders," which details the week-to-week interaction of Quant UXRs with colleagues on a product development team. For readers who are new to industry, this chapter will help you understand how to influence decisions in product teams. For more experienced practitioners, our advice will assist you to conceptualize, diagnose, and avoid pitfalls with stakeholders.

Chapter 14, "Career Development for Quant UX Researchers," extends the time horizon and reviews individual development over time. We highlight the importance of considering personal styles and goals when making career choices, and give descriptions of very senior career levels that are rarely discussed.

Finally, we close with Chapter 15, "Future Directions for Quant UX," where we share our expectations for the likely paths that Quant UX may take over the next decade.

CHAPTER 11

UX Organizations

In Chapter 2, "User Experience and UX Research," we described the relationship between UX organizations and other parts of technology companies. Now we will examine how UX organizations themselves are structured and common models for positioning Quant UX within them. There are a couple of patterns that often work well, in our opinion, and others that do not work well. In this chapter, we outline the arguments for and against each model.

The organizational models here apply to large organizations, such as big tech companies. A startup with 15 employees cannot have such a complex structure. Even if you work in another kind of organization, you may be interested for various reasons: your organization may grow; you might change jobs; some of the pitfalls described here also occur in other models; or because you do consulting work with a large organization, and understanding its operations will benefit your work.

The models we discuss here are not intended to be perfectly descriptive or exhaustive; there are as many variations as there are UX organizations. However, the distinctions, advantages, and pitfalls that we describe occur across all of those variations. The discussion here are valuable even when aspects of the models are combined in different ways.

11.1 Typical UX Organization Models

As described in Chapter 2, UX teams typically comprise UX designers and UX researchers, potentially with other roles such as UX writer, UX producer, and UX product manager (PM), among others.

Two common models for organizing a UX group are *role-centric* organization and *product-centric* organization. In the role-centric model, designers report to a design manager, while researchers report to a research manager, and similarly for other roles.

© Chris Chapman and Kerry Rodden 2023
C. Chapman and K. Rodden, *Quantitative User Experience Research*,
https://doi.org/10.1007/978-1-4842-9268-6_11

There may be multiple design or research managers, and all of them report to a UX director (who might have other titles such as creative director or VP of UX).

In the product-centric model, UXers work in pods with multiple UX roles represented. All members of the pod work on the same product. The pod may include a few designers and a researcher or two, alongside other roles such as program managers, writers, or front-end engineers.

Although we discuss them separately, the models in this chapter may overlap in practice. For example, a role-centric team may assign out researchers to work on product-centric, multidisciplinary pods. In that case, the advantages and disadvantages we discuss here would apply according to where the organizational *power* lies—the underlying structure that most closely aligns to the organization's patterns for hiring, product decisions, performance reviews, recognition, and so forth.

To take such combinations to an extreme, the most complex possibility would be a *matrixed organization* where Quant UXRs report to multiple managers with differing areas of responsibility. For example, a Quant UXR might report to a Quant manager for research expertise; to a UX design manager for contributions to user experience; to a product manager with responsibility for product decisions; and so forth. In practice, we have never seen such a model for UXRs, and we would worry greatly about the potential for conflicting goals.

11.1.1 Role-Centric Organization

In the role-centric model, each job function reports to a manager with the same background. For example, UXRs report to a UXR manager. This is not unique to UX roles; it is the single most common model across many industries. Staff attorneys and accountants report to partner attorneys and accountants; soldiers and ensigns report to more senior officers; software developers report to engineering managers (who are developers); retail workers report to a department or floor supervisor. A typical role-centric UX organization chart is shown in Figure 11-1.

The primary advantage of this model for UX researchers is that managers and their reports are deeply familiar with the UXR role. One may expect a manager to deeply understand the technical requirements and expectations of the role, to advocate strongly for it, to understand contributions when it is time for performance reviews, and to have in-depth experience with organizational demands such as budgeting and hiring. Similarly, the contributors on the team understand the background of their manager, and—in an ideal case—learn technical skills from them.

However, there are less obvious disadvantages. First, the apparent closeness of the role may obscure important differences in background, skills, or expectations. No research manager can be an expert in everything, and there is a risk that managers will have an overly narrow view of the field, overestimate their own expertise, or assume that their direct reports are more similar to themselves than they really are. A manager with a different background might be more open or inquisitive.

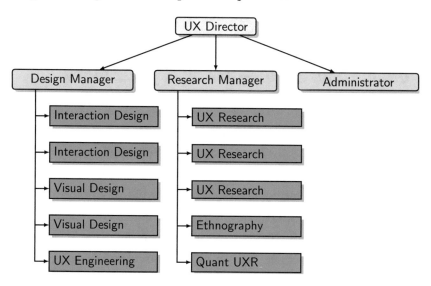

Figure 11-1. *Typical role-centric UX organization, where Quant UX resides in a larger UX research team supporting multiple products (or multiple areas of one product)*

Second, in our view, the challenges confronted by UXRs do not typically require *research* expertise from a manager. Instead, they require *organizational* agility and influence, such as knowing which research projects will be most likely to have impact at a given point in time. When managers come from same role, they may lose sight of organizational issues and spend too much time on technical aspects of the role.

A third concern is specific to research teams: it is tempting for a research team to spend time on *foundational* work that explores a problem or strategic need in depth. For example, a research team might engage in an extended project to deeply understand and describe the needs and behaviors of user segments or personas; or a longitudinal project to engage repeatedly with a group of users to track their needs and product usage; or do a comprehensive review of behavioral metrics and set up tracking scorecards for a product.

You might think, "Those projects sound great! I'd love to do that." And you're right—they do sound great, and researchers do love them. Our caution is that research teams should not undertake such projects without a clear connection back to how they will have an impact on the product, such as informing *decisions* to be made. It is easy for UXR teams to launch projects that appeal to themselves and hope that somehow they will later influence product decisions. That is a huge risk. (By the way, a similar process sometimes occurs with UX *design* teams who spend too much effort on design frameworks or speculative product visions.)

11.1.1.1 Quant UXRs in Role-Centric Organizations

In role-centric organizations, Quant UXRs are generally part of an overall UXR team reporting alongside general UXRs. In the case of very large organizations, there might be a separate Quant UXR team with its own manager, where all the researchers are Quant UXRs.

The advantages and disadvantages for Quant UXRs in this situation are intensified versions of our general observations just described. Quant UXRs will find a higher degree of understanding of their work in such a situation. Also, it will be easier to pair with a general UXR to do projects that combine qualitative and quantitative aspects.

However, on a UXR team, Quant UXRs may find that their contributions are not understood as well as those of other UXRs. General UXRs often work directly with UX designers, whereas Quant UXRs more frequently work with PMs or executive stakeholders. Those differences may lead to confusion among UXR managers and difficulty assessing the contributions of Quant UXRs as compared to general UXRs. General UXRs often recommend direct changes to products based on research such as usability lab studies or field work. It may be relatively easier for them to demonstrate "impact" on the product (see Section 14.3.2) and to have that impact understood by management. Quant UXRs are less likely than general UXRs to have management understand their roles, and therefore will have less benefit in this model.

Depending on the performance review model of the larger organization, such as whether it performs stack ranking of employees (see Section 14.3.1), there may also be competition among researchers. This may lead to nonproductive activity such as filing excessive or frivolous reports, or running extraneous studies to pad one's statistics. In the worst case, it may lead to hostile engagement such as criticizing colleagues' reports in an attempt to boost oneself.

11.1.1.2 Notes on Success with Role-Centric Organizations

Our overall take on the role-centric UXR model is that it is a viable model, although not an ideal one. From a management and organizational perspective, the key to success is close alignment with important product goals, taking great care that the team is delivering important results that affect decisions for the product, and not merely nice-to-know findings that may have little effect.

At the same time, it is important to establish long-term plans with the research team. It takes significant time—starting with roughly 1 month for simple projects, and often 3–6 months for complex projects—to plan, conduct, and report high-quality research. A team manager should establish those expectations with upper management and horizontal stakeholders and protect the UXR team from constantly changing priorities.

From an individual Quant UXR point of view, a key aspect is to ensure that your projects are aligned with important product decisions. It is risky to engage in long-term technical projects without close communication and check-ins with the product team. In this kind of model, we recommend that you work closely with at least one general UXR on the team. Such a partnership will have higher engagement with stakeholders, simply because there are two or more of you, and you will have a trusted partner to provide feedback, both on the work itself and for organizational matters (such as performance reviews or budget requests). Even more importantly, when research involves multiple perspectives and methods, it is better research.

Put simply, good research takes time and collaboration. Organizations need to understand that UXRs are not utilized well and will not be happy as an "on-call" service.

11.1.2 Product-Centric Organization

In the product-centric model (sometimes described as a "UX pod" approach), the lowest-level UX teams comprise diverse skills that are needed for product development. For example, a UX team may have one or two designers and researchers, perhaps along with a writer, program manager, or other UX role. This group is managed by a UX manager, who may have background in design or research. Day to day, the team works closely with one or two PMs or engineers who are responsible for product decisions and delivery. A typical product-centric organization chart is shown in Figure 11-2.

The great advantage of this model is that all UX work is closely connected across roles and directly engaged with product development. UXRs will hear research questions from their design colleagues and PMs, while daily engagement ensures that the team will engage with research, observe user sessions, and use research findings to make product decisions. Thus, this model corresponds well to the ultimate goal of a product development organization, namely, developing products.

Overall this is our favorite model. We believe it has the highest odds of positive product impact (from the integrated team), yields the fastest velocity (from close daily engagement), and is likely to lead to the greatest happiness among team members (because of the reinforcing roles, shared purpose, and the fact that each person's success comes from noncompetitive contribution to the success of the whole team).

However, no model is perfect—because people (and products, customers, and organizations) are not perfect—and there are common problems that arise in this model. First, the different roles may have inconsistent expectations that are difficult to resolve in the daily peer-to-peer model. For example, a designer may expect that research will perform iterative usability studies that examine successive changes in a product, whereas research may believe that such work is unnecessary, overly detailed, or would not meet standards for quality research (such as the quality or size of a readily available user sample).

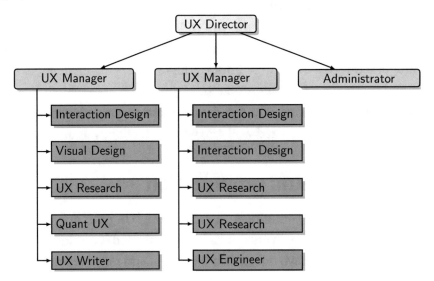

Figure 11-2. *Typical product-centric (UX pod) organization, where Quant UX resides in multidisciplinary UX teams that support single products (or single areas within a product). This is often a desirable model because Quant UXRs are closely engaged with design, product engineering, and decisions*

Second, the team members will differ in experience and status, and this may lead to conflicting goals. Consider the previous example, where a designer seeks immediate feedback on a design. If the researcher is relatively senior in status, then they would likely much prefer to manage larger scale, longer term projects. Yet the design colleagues may not be able to benefit from research that involves higher-level product strategies.

Third, the close team engagement in a UX pod almost inevitably leads to a high commitment to product success. Why is that a problem? Because it is easy in research findings to focus on the potential positive outcomes and downplay data that suggest negative results. Consider again the case of iterative design research, and suppose that there are iterative usability studies that demonstrate interaction concerns with the product. Following each study, a researcher may present recommendations for improving the design.

However, a more important finding could be that research sees a consistent pattern of user difficulty and failure that suggests limited value of the design or product. That would be a difficult message to communicate in a close-knit team, and might be viewed as a betrayal. It is much easier for a researcher to avoid such confrontation—or the imagined risk of such confrontation—and instead to emphasize more "positive" recommendations.

A final consideration is that a contributor in a UX pod may be a "singleton" who is the only person in the pod with a given role. They may have no close technical colleagues (such as other Quant UXRs). In our view, whether that is a disadvantage or an advantage varies according to the person and situation. For a relative newcomer to UX, it may be a disadvantage. For a more senior contributor, or a junior one who wants more autonomy and responsibility, it may be an advantage. In either case, we strongly encourage Quant UXRs to develop acquaintances with others in their role across the larger organization or industry.

11.1.2.1 Quant UXRs in Product-Centric Organizations

For a Quant UXR, the success of such a role is likely to depend heavily on the degree to which the other team members understand Quant UX—or at least understand its value and are willing to embrace the Quant UXR's expertise. If other team members expect the Quant UXR role to be one of immediately answering questions through data queries and the like, it is likely to be immensely dissatisfying for a skilled researcher.

On the other hand, the UX pod organization may be one of the most highly rewarding situations for a Quant UXR. It can bring a close interaction between the decision-makers (the team) and the research questions. There may be the opportunity to work closely with a general UX researcher, with qualitative and quantitative projects that complement one another and deepen the team's understanding of users and product solutions. The rapid pace of pod research and design means that a UX researcher's work will be directly seen in the product.

Also, for Quant UXRs who are relatively new to UX, or who have worked in larger, perhaps centralized teams, the pod structure offers the opportunity to understand product development and other roles in depth. In such a role, you will interact directly with PMs, designers, engineers, and qualitative UX researchers, and will come to understand their roles in much greater depth than in other organizational models. This experience will greatly benefit you over the span of a career.

11.1.2.2 Notes on Success with Product-Centric Organizations

The product-centric organization is our most-favored structure for UX teams. However, it is not guaranteed to succeed and we have a few recommendations for team members and managers.

First of all, this model depends heavily on the interaction style of the members, who need to be professional, autonomous, and relatively expert in their areas. They must respect the expertise of others and afford ample autonomy to their colleagues. Thus, it is not an appropriate structure for placement of domineering, disrespectful, or self-absorbed employees, nor those who overestimate their expertise in others' areas. We recommend that managers carefully consider the interaction styles of team members before placing them into such pods. The members should be as honest with themselves as possible about their own styles. Ideally, such a team might be largely self-formed, as individuals who demonstrate collaboration and complementary styles are invited to work together.

A key challenge in the UX pod model is to watch for potential boredom or stagnation. High performers, especially senior contributors, will want opportunity for larger-scale, possibly higher-risk projects. One possibility may be to rotate the team to such projects over time, such as inviting the entire team to take on a large new challenge.

Another issue will arise as the team changes over time, perhaps through attrition, diminished need after a product ships, or by addition of members through growth. New members may not fit as well in an established team, and longer-term members may dislike the changes. It would be unusual for a team to persist for more than 2 years or so, given the usual timing of product releases, strategy changes, hiring, promotion, and turnover.

For Quant UXRs on such teams, the most important element of success is to learn about colleagues' work and build relationships that ensure that you have interesting and important research questions. This will lead to research projects that influence decisions. Keep research simple—at least in the reports and deliverables—and do not attempt to impress colleagues with technical difficulty. Before sharing results with stakeholders, give previews to team members. Their feedback will help you avoid surprises or misunderstandings when the results are presented more widely.

11.2 Other Organizational Models for Quant UXRs

The two models previously described—role-centric UX research teams and multidisciplinary UX pods—are most common in our experience. In this section, we discuss two alternative models that focus on analytic skills. These place Quant UX research either in a team of its own or in a larger analytics team.

11.2.1 Centralized Quant UX Research Teams

A centralized Quant UX research team is similar to the role-centric UX research teams we considered in Section 11.1.1. There are two modifications: the team comprises Quant UXRs instead of a mix of UX researchers, and the team works across multiple products or all products for an entire firm. This model presents intensified versions of the advantages and challenges we described for general research teams. A typical centralized organization chart is shown in Figure 11-3.

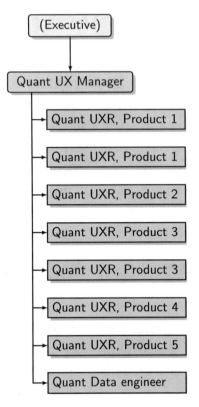

Figure 11-3. *A centralized Quant UX organization, supporting multiple products. This model offers opportunities for strong technical collaboration but may have difficulty aligning with the needs of the larger product development organization*

The benefits of this model are the high degree of support, understanding, and interaction among Quant UXRs. The team can develop its own culture and evolve itself over time in alignment with that culture. This will positively shape the expectations for the products that the team works on: there will be consistent framing of Quant UXR problems and questions, and greater ability to select problems that best fit Quant UXR. Likewise, team members will have greater flexibility to select products or problems of interest over time. The diversity of skills on the team will provide expertise across a large set of research needs.

In a centralized model, demand for Quant UX research across the product space is likely to be vastly larger than the team can support. The team will be able to select the most important and rewarding challenges, leading to higher team satisfaction. It will also be easy to demonstrate the need to grow.

The core problem in this model is that Quant UXR is disconnected from the needs of individual products and from UX research more generally. There are likely to be continual issues around deciding which products to support and how to align the team's skills and effort relative to product needs. This places a lot of pressure on the manager, who needs to understand priority and likelihood of project success across a broad range of product areas and stakeholders, and may have limited ability to coach their team members on matters relating to organizational context. Some products or projects will be much more appealing to team members than others, and some may offer a greater degree of apparent "success" that is mostly unrelated to the Quant UXR's contribution.

Because this model separates Quant UXRs from general UXRs, it may end up emphasizing the *Quant* aspect of the role more than the *UX* and *researcher* aspects. It pulls for Quant UXRs to differentiate from one another by developing greater and greater technical expertise within their own depth skill areas. That may be rewarding and appealing to team members, but will it yield the greatest flexibility and long-term success for them and for the larger organization? As we have noted repeatedly in this book, we prefer to think of Quant UXRs as *UXers first*, who bring research skills to UX problems, and happen—lastly—to have special depth in Quant methods.

11.2.1.1 Recommendations for Centralized Quant UX Research Teams

Kerry managed the first Quant UXRs at Google under this model; it can be appropriate in situations where Quant UX research is new to a firm, as it will assist with building Quant UXR expectations and meeting initial demand while simultaneously evolving the role to better fit a firm's needs. Similarly, it may be a good model when many Quant UXRs are new to the role or industry, because it offers strong opportunities for close mentoring, acculturation, and learning skills with support from colleagues.

We believe this model should be reevaluated over time. It may make sense at some point to break up a central team and distribute its members to individual product areas, especially when they become more strongly aligned with UX generally (see Section 11.1.2).

Managing a Centralized Quant Team

When managing a centralized Quant UXR team, it is crucial to address how research bandwidth (i.e., headcount and person time) is allocated to the competing demands from various product teams. We recommend developing a clear set of general guidelines with upper management, and a *lightweight* process for selecting projects and reviewing allocation.

247

We emphasize "lightweight" because it is often appealing to develop a more complex process that gathers requests from the organization (perhaps through an "intake" form). Unfortunately, despite the good intentions of such an intake process, we have found it often to be detrimental. It implicitly communicates to teams that Quant UXR is an on-demand service. Because requesters are unlikely to understand the value and structure of larger research projects, and yet have immediate needs for simple data, the intake process often focuses a team on routine analytics rather than vital but longer-term research.

A better process is to engage directly with stakeholders to understand their needs and then use the team's expertise in research design (while considering the members' set of unique skills) to recommend research projects. Depending on the size of the organization, this can be very demanding for the Quant UXR manager and ideally should be part of a broader UX research prioritization process, both for consistency and to ensure that project assignments are considering the partnerships between Quant UXRs and general UXRs.

We also recommend that each team member be assigned to no more than one product area at a time. In particular, such a team should avoid a matrix model where multiple Quant UXRs are each assigned to multiple products in an attempt to mix their skills. This is for both technical and interpersonal reasons. On the technical side, every product poses unique issues that a Quant UXR needs to master, such as how its product logs are structured, details of its code base or user interface, and how to reach users for feedback. On the interpersonal side, there will be many team members—PMs, designers, general UX researchers, engineers, testers, writers, executives, and others—with whom the Quant UXR interacts. When Quant UXRs are allocated to more than one product area, they are not able to develop adequate engagement.

A reasonable minimum duration for assignment to a team depends on the amount of technical ramp-up required (for example, the complexity of the product's logging infrastructure), but should be ideally at least 6 months. There is no particular maximum, except to consider whether rotation might benefit both UXRs by diversifying their experience and the product by refreshing the available analytic approaches.

Recommendations to Individual Quant UXRs

For Quant UXR team members in a centralized organization, we have two recommendations. First, do not assume that this model will be permanent. Build connections with colleagues who are situated inside product teams, both teams with which you work as well as others across the organization. Make connections especially with general UXRs and UX managers. If your team is disbanded, or if you simply wish to make a change, these connections will help you consider your options of what to do next—and might even offer you a position.

Our second recommendation is related: learn as much as you can about how general UX research operates. If you later join a general UX research team or a UX pod, this knowledge will be beneficial. You might even decide to join an organization that does not have specific Quant UX research positions, in which case you could consider general UXR positions. Besides partnering with general UXRs when possible, we have a more immediate suggestion: ask them whether you may observe qualitative research sessions such as usability lab studies. Ask questions about the methods and how they arrive at results. Ideally, after some observational experience, consider running your own qualitative study (see Section 14.5.1).

11.2.2 Quant UX in a Data Science or Analytics Team

The second skills-oriented team model occurs when Quant UXRs are part of a dedicated analytics team. Such a team may also include data scientists, business analysts, product analysts, market analysts, marketing researchers, and similar roles. An example of an analytics team organization chart is shown in Figure 11-4.

We believe such an organization is *not* a good choice for Quant UXRs. UX researchers will be unlikely to succeed or thrive in such a situation, and the team itself is likely to have an unstable relationship to the larger organization. At best, it will become a pure data science or business analysis team.

Such an organization may seem appealing because such roles often share technical methods and may be viewed as complementary in their approaches. For example, both Quant UXRs and data scientists may perform statistical modeling of product logs. One might view the Quant UXRs as focusing on user needs, while data scientists focus on the business needs.

The problem is that this emphasizes a superficial similarity, namely that all of these roles access data and perform some kind of analytics or statistical modeling. In our view, this is similar to saying that accounting and software development should be on one team because both roles use math and keyboards! As we note in Section 3.4, business analysts and data scientists are unlikely to have depth in human behavioral research and related experimental design skills. They tackle different kinds of questions than UX researchers, and deliver insights to different groups of stakeholders.

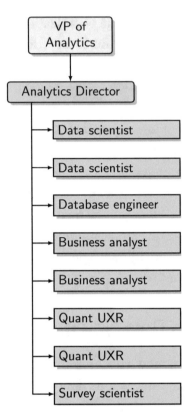

Figure 11-4. *A typical analytics organization, with Quant UXR as one specialty. This model is usually not a good choice for Quant UXRs because it does not emphasize the human-centered approaches of UX research*

11.2.2.1 Recommendations for Quant UX in a Data Science or Analytics Team

We recommend that Quant UX research not be placed in a data science or business analytics team. The roles differ too much in background, expectation, experience, skills, and paths for organizational influence. Such a structure is unlikely to be stable over the long run.

However, in the short run, this might be viewed as a temporary superset of a centralized Quant UX team (see Section 11.2.1). With that in mind, we recommend to follow the suggestions there to build UX relationships and connections, and to consider—or even better, to *plan*—how to eventually disaggregate the team and move Quant UXR directly into close alignment with general UX research. For a team manager, it is a valuable outcome to use the success of an initial central team to later establish more specific and product-connected offshoots within product teams.

For individual Quant UXRs on an analytics team, it may be viewed as an opportunity to consider whether you wish instead to switch to a data science or business analytics career. Those are excellent career choices that emphasize different skills than Quant UX research. However, if you wish to remain in Quant UX, we recommend to switch teams when a good opportunity arises, and meanwhile to develop connections and skills as noted earlier for centralized teams.

11.3 Advice for Managers of Quant UXRs

For managers of Quant UXRs, we have a few observations to add to our suggestions in the previous detailed sections.

The most important one is this: if you do not have any quantitative background yourself, *do not worry about your ability to manage Quant UXRs*! As we emphasize throughout this book, Quant UXRs are a variety of UXRs and UXers generally. We often talk with UX managers who are concerned whether their skills are appropriate to manage Quant UXRs, and we always say, "Yes!"

More specifically, Quant UXRs are able to get technical help from one another, colleagues in engineering or data science, and online sources. What they need from management is assistance with identifying key decision points, finding stakeholders who are great partners, communicating results, and being shielded from random, disruptive, and immediate but low-value work.

In the following sections, we discuss additional ways that managers can help their Quant UXRs. For individual Quant UXRs, these sections may help you to set expectations and better advocate for yourselves with management.

11.3.1 Access to Stakeholders and Data

In large organizations, Quant UXRs often have difficulty in accessing data and executive stakeholders. A good UX manager will assist with these issues.

Stakeholder access is crucial at four points: clarifying the research problem; getting input on near-final answers to the question; seeing and understanding the final answer; and providing feedback on the impact, such as comments on a Quant UXR's performance review. Too often research becomes a "shelf report"—a report that is delivered but has no value because it is filed away. This often happens because of some disconnect between the Quant UXR and the stakeholder, such as lack of Quant UXR involvement in clarifying the research question, or lack of Quant UXR consideration of stakeholder needs when reporting findings. (We say more about this issue in Section 13.2.1.)

A UX manager can assist a Quant UXR to gain access to executives to ask questions and set expectations about the research process. We recommend to communicate clearly what will be required from stakeholders and what they may expect. For example, a UX manager might say, "My initial guess is that this will take about 2 months. We'll schedule time to detail your questions and how you will use the results. We'll follow up 4 weeks later with preliminary results to see what additional questions you have. Then I expect a final answer about 2 weeks later."

At this point (if not earlier), the process might be given to the Quant UXR to manage. Let the Quant UXR schedule and run meetings with executives and always let them present the results. The manager might follow up immediately with the executive to solicit feedback—if positive, this can be conveyed to the Quant UXR and filed away until performance review time, and if negative, the issues can be addressed while still recent.

Closely related to this is access to data. The twin problems of data for Quant UXRs are no data and data that can't be accessed. An *absence of data* may reflect organizational issues: collection of data such as product logs requires work from other disciplines such as software engineering, but those roles may assign low priority to such work. Instead, they devote their time to areas where they are rewarded, which is developing and shipping product features. A UX manager can gather support from executives for the importance of collecting the data and engage in problem-solving with other roles. For example, if the problem with software engineering is a lack of time among their team, one possible solution might be for a Quant UXR to take on some of the engineering work. (Be sure to talk with your Quant UXR before offering this in order to assess their skills and interest in such work.)

Data access is often limited by engineering systems' permissions as well as corporate policies about privacy, data storage, data retention, and the like. It is not unusual for data to exist that Quant UXRs would like to access, but they do not have the necessary permissions. You might say, "Well, give them permission!" and we would answer that it is complicated. Good analytics often start with data exploration, and a Quant UXR might not know in advance exactly what they need to access. A blanket permission of "access to everything" is excessive and may be explicitly denied or even impossible (e.g., when data are kept in multiple systems with intentionally de-identified keys that cannot be joined).

Solutions to these problems vary widely across organizations, products, and questions. We do not offer a prescription except for this: a good manager can help their Quant UXRs with data access. First, a manager should understand the situation and clarify expectations with stakeholders. Second, a manager should take on as much of the organizational work as possible—such as communicating the importance of data access—so Quant UXRs may focus their time on research.

11.3.2 Shield from Immediate Requests

Good research requires time to think, explore solutions, code and revise code, iterate on analyses and how to present results, engage with stakeholders along the way, and present results at the end. One of the most destructive (although well-intentioned) ways that managers engage with research is to short-circuit that process with immediate requests.

Here is a common way that happens. The UX manager meets with an executive—let's say, a vice president—who asks a question about users. Perhaps the VP question is, "Does usage of feature A lead to more engagement with feature B?" The manager brings this question to the Quant UXR and asks them to explore the question, explaining that this is a "huge opportunity" for impact and visibility because it is being asked by a VP. In this situation Quant UXRs have almost no choice other than to agree to the request.

When we reflect for a moment, we observe several things about this situation. First, the VP's question may be idle speculation, and they likely do not understand the amount of work that is needed to answer it. Second, there is no clarity as to why the question is important or what decision it might inform. Thus, any research at this stage would be unaware of underlying assumptions or business needs.

Third, the concepts and assumptions—what is "usage" (and who counts as a "user"), what is "engagement," what is meant by "feature A" and "feature B," what is meant by "leads to...," and what level of effect size would count—are all unspecified. Fourth, there is no consideration of what other work will *not* be done (the *opportunity cost*) if the Quant UXR takes on this question. How high is the priority for this request?

Finally, it is purely speculative whether an answer to the question will have any impact. Not only do we not know its importance, we don't know how a negative result would be interpreted. Suppose the result is "no effect, as far as we can tell." Will that be perceived as a waste of time? As incorrect? As an implicit challenge to the VP? As a sign that the Quant UXR is incompetent?

A good Quant UXR takes seriously the need to establish data definitions, problem clarification, requirements to establish a causal link, and so forth. This may lead to an executive's apparently straightforward question becoming a large project. Yet, when a result is promised without upfront clarification of the scope, priority, or actions that might be taken, the likely result will be blockage, failure, or disappointment of the Quant UXR, the manager, or the VP. In short, the expected outcome is a waste of time, disruption of projects, and damaged morale of the Quant UXR. This outcome is known as *randomization*.

What should a manager do instead? Avoid committing to answering such requests and instead ask for clarification of their importance. Is it truly important? How will the answer be used? The manager should also communicate the trade-offs and potential outcomes. Clarify the expectations about how good research gets done. Suppose it takes a month to get an answer; is that OK? Is it worth hiring someone else? Should their team stop work on other projects in order to answer this?

Often, executives will respond that the question is not very important and drop the question. At other times, they will learn something important about the underlying challenges. This not only limits the immediate damage, it has the positive effect of setting the team up for more success on its own terms. File away the question and consider how you might pleasantly surprise the executive later by being prepared to answer it or similar requests after evaluating the need and working through details at a sustainable pace.

What if upper management strongly believes that Quant UXRs should be answering such questions immediately, querying available data? That would be a sign that a Quant UXR team may be the wrong choice; a business analytics team may be more appropriate than a team of researchers (see Section 3.4).

11.3.3 Growth Opportunity

Because it is a relatively new career area with few established paths, Quant UXRs and their managers may be uncertain about career opportunities. This is an area for discussion and tailoring to every unique Quant UXR, and we have more to say about it in Chapter 14, "Career Development for Quant UX Researchers." In the present context of organizational structure, we have a few general thoughts.

First of all, look for opportunities for increasing *larger and longer-range projects*. It is often helpful to think of quant projects as inverted stacks, where foundational work (perhaps logs engineering, A/B tests, or a survey infrastructure) enable an increasingly broad range of later applications and analyses. For example, causal modeling might be applied to logs; multivariate testing might be developed on top of a solid experimentation platform; and segmentation, market preference, and other analyses developed in successive waves of survey work.

The longer time horizon may work well for Quant UXRs to have ownership and day-to-day autonomy. (Note that a corollary of this suggestion is to be very cautious before committing to such "higher stack" work when foundations have not been laid. For instance, it is risky to jump immediately into a segmentation project without prior survey foundations.)

Second, as we noted earlier in this chapter, Quant UXRs should do as much as possible with regard to *stakeholder engagement and presentation* (see Chapter 13, "Research Processes, Reporting, and Stakeholders"). We have too often seen Quant UXRs sit on the side while others present their work. Even when that is due to personal preference, we encourage working to develop more stakeholder engagement. Besides bringing higher fidelity to user data and analyses—which the Quant UXR knows best of anyone—this leads to higher-quality *questions* for future work. And it helps with visibility for performance reviews and promotion.

Third, we encourage strong *community engagement*, either across a firm or externally (or both). In particular, we view *teaching* and *publishing* as being exceptionally high-value activities that are too often overlooked or discouraged. For a Quant UXR, teaching and publishing will lead to a higher level of understanding and knowledge. They may also be emotionally rewarding.

What is less obvious is that teaching and publishing bring enormous benefits to firms and product groups. When UXRs elevate their skills, that leads to higher-quality work and more-accurate product decisions. They will bring new skills, and are more likely to find any errors or suboptimal practices in existing work. Positive reputation, among internal colleagues or external audiences, assists a team to recruit talent. Exposure to outside ideas—through reviewer feedback or attending conferences—will lead them to learn about new approaches, techniques, and possibilities.

A common question is this: "How can we publish? Our data are proprietary, and we don't want to reveal our methods to competitors." There are two answers to this. First, many conferences and journals are accustomed to this and will allow data to be disguised. In our experience, conferences are particularly eager to have industry engagement and will be flexible. Second, few methods are truly revolutionary, and we believe the risk of disclosing them is almost always lower than the benefits to be gained from feedback, peer review, and cross-pollination with others. Also, if your Quant UXR is ahead of the competition, then they will be *better* able to benefit from such engagement. Because they are starting out ahead, any acceleration will be a net positive to your organization. In short, there should be a strong presumption that teaching and publishing will have a net positive contribution, both to individual Quant UXR satisfaction and to an organization's proficiency and user centricity.

11.3.4 Help with Determining Impact

A common concern for Quant UXRs is how to demonstrate "impact" for a product. Impact is a nebulous term for the degree to which a contributor influences a product's design, delivery, market position, or success. For *general* UXRs, impact often occurs through collaboration with UX design. The UXR provides user feedback on a design; the designer alters the product UI; and thus the UXR has "impact."

For Quant UXRs, a similar process sometimes occur, yet more often the process is diffuse and the parties who are influenced have less direct responsibility for product implementation than designers. For example, a Quant UXR might provide evidence to a PM that users are showing decreasing engagement with the product. The PM talks with colleague PMs, designers, engineers, and executives about the trend. After some time, new features are launched with a hope to increase engagement. By that time, there will have been many sources of input and the Quant UXR's contribution may be forgotten. Did the Quant UXR have impact? Will it be recognized and supported by the PMs and others?

A good manager can help Quant UXRs to demonstrate impact in several ways. First, the manager should help with the selection of projects to begin with (see Section 11.3.2). Another way is to periodically poll Quant UXRs' partners about their contributions, especially shortly after project delivery. It may be awkward for a Quant UXR to ask a PM, "Would you write something now that I can put in my review several months away?" But that is a perfectly reasonable—indeed, almost expected—request from a UX manager.

We say more about impact with regard to performance reviews in Section 14.3.1.

11.3.5 Stay Out of the Way

Our final recommendation to managers is to trust your Quant UXRs and let them do their jobs. We have observed a few areas where UX managers often, and we suspect unknowingly, create problems for Quant UXRs.

One problem is posing hypothetical, spur-of-the-moment questions. We described this in Section 11.3.2. When seeing results, it is natural to wonder, "Could this be due to X?" and ask that the Quant UXR take a look at the relationship. This is a bad idea. It will randomize the Quant UXR, take much more time than the manager expects, and usually produce a null result or, worse, a spurious result. Resist this! If the question is truly important, treat it as a new research inquiry and prioritize it against other work. If it is not truly important, don't ask.

Another common problem is micromanaging areas where the UX Manager feels relative confidence. Survey design is particularly prone to this. Surveys appear to be nothing terribly special; surely anyone could write a good one! And anyone could suggest wording and new questions, right? Resist this as well. Instead, advocate for qualitative pre-testing of surveys with users. This will improve a survey from the users' point of view rather than managerial opinion.

Finally, resist dictating research plans, technical approaches, or analytic methods. Let the Quant UXR develop plans and analyses. At the same time, show interest in the methods. Ask how the analyses work and clarify your understanding of them. That will make your work as a UX manager more effective and the Quant UXR will appreciate your interest.

11.4 Key Points

We considered several organizational patterns for Quant UX research and their advantages and disadvantages. Although the structure of a UX team may be dictated by the larger organization, there are ways for Quant UXRs to form closer product partnerships, to be more effective, and to be happier. In the best case, a UX or research manager will be able to implement one of the structures we recommend.

Table 11-1 summarizes pros and cons of each model. Although there is no perfect model, we believe the most effective model in general is the product-centric UX pod approach. If an organization uses a more traditional role-centric model, it is important to consider carefully how to show Quant UXR impact on the product and provide technical mentorship and peer support.

Table 11-1. *Advantages and Disadvantages of Common Quant UXR Organizational Models*

Structure: Quant UXR in…	Key Advantage	Key Disadvantage
Role-centric team (traditional)	Close engagement across UX research	Difficulty showing impact on product
Product-centric team, multidisciplinary (UX pod)	Clear research questions and product impact	Difficulty managing pressure from the team
Centralized Quant UXR team	Close technical collaboration and mentoring	Disconnection from other UX research and product teams
Data science or analytics	Meets superficial expectation of technical similarity	May become neither research nor UX

In any model, we recommend that Quant UXRs form close partnerships with general UX researchers, with product managers who bring them interesting and important product questions, and with other Quant UXRs across the organization (or the industry at large). These partnerships will help Quant UXRs to avoid being frustrated by the limitations of any particular organizational model.

11.5 Learning More

We discussed UX organizations from the point of view of Quant UX researchers, and primarily with respect to large firms. In *User Experience Management* [86], Lund provides guidance on all aspects of UX organizations and UX management, for all sizes of firms, across all UX disciplines. If you are managing a UX team, or considering a move into management, it is recommended reading.

A useful complement to Lund [86] is Larson's *An Elegant Puzzle* [79], which describes models, approaches, and best practices for managing software engineering teams. Although it discusses software engineering, it is directly applicable to UX organizations as well. Also, it will help you to understand the workings and dynamics of your product engineering counterparts.

CHAPTER 12

Interviews and Job Postings

In each of our chapters on Quant UXR skills (Chapter 4, "UX Research," Chapter 5, "Statistics," and Chapter 6, "Programming"), we described common interview topics. Here we review common models for Quant UXR interviews at a more general level, discussing who participates, what happens before and after interviews, and how you might prepare (and *not* prepare) for them. We also discuss job postings and share our thoughts about warning signs that may appear in them.

Hiring processes vary dramatically according to organization size and structure; the practices of a large firm like Google or Microsoft are more complex than those of a startup. This does not mean that large companies are necessarily slower, because their structures might either speed up the hiring process or slow it down.

For purposes of discussion here, we will describe the processes we've seen in *large* to *very large* organizations. In smaller firms, the steps are likely to be similar but less formal, with fewer people involved. As we noted in Chapter 1, "Getting Started," we write from the perspective of large tech companies hiring for permanent, full-time roles. Other roles such as temporary, contract Quant UXR positions also exist. Much of the discussion applies to those roles as well, although the hiring processes will be conducted differently.

12.1 General Quant UXR Interview Process

From a candidate's perspective, the interview process begins long before you decide to apply. In Section 12.4, we discuss how you might make such a decision. Here we will examine the process from the firm's perspective.

When reviewing an applicant, an organization looks to answer the following questions in successive order:

© Chris Chapman and Kerry Rodden 2023
C. Chapman and K. Rodden, *Quantitative User Experience Research*,
https://doi.org/10.1007/978-1-4842-9268-6_12

1. *Does the candidate look potentially suitable for the position?* This
    is determined by some combination of keyword recognition
    applied to a résumé and a *sourcing recruiter* reading a résumé or
    other documents. The sourcing recruiter is a human resources
    (HR) specialist in charge of finding candidates and seeing them
    through the initial stages of an application. At this stage, your job
    is to get the right keywords onto your résumé while not overstating
    your competence (interviews might ask you to demonstrate
    anything you claim). The best source for keywords is in the job
    postings themselves; incorporate as many of the stated skills and
    requirements as you can while being accurate.

2. *Is the candidate in the top N we would consider?* An organization
    is unable to interview everyone and needs to select a smaller set
    of the most promising candidates. The set of candidates to be
    interviewed will be determined by recommendations from the
    sourcing recruiter and reviews of candidates by hiring managers
    (HMs). For example, the recruiter may send selected résumés to
    HMs and ask them, "Are you interested in this candidate?" There's
    not much you can do about this stage because it depends on who
    else is applying and exactly what the HMs are seeking.

3. *Should we invest in a full interview with the candidate?* Interviews
    are costly in time, effort, and (sometimes) travel costs. Before
    investing in that, companies typically conduct one or two phone
    interviews to assess your skills, interests, and potential fit. A *phone
    screen* tries to answer the question, "Is there a good chance this
    candidate can succeed during a full interview?" and it answers
    that by briefly sampling your skills in the technical areas described
    in Chapters 4–6, along with your communication style. Often a
    first phone interview is conducted by the sourcing recruiter, and a
    second one involves a technical interviewer (see Section 12.3.1.)
    For you, the best preparation is to ensure that you will be mentally
    fresh for the call. Set aside 30-60 minutes of unscheduled time
    before the call so you will be relaxed. Arrange to be in a quiet
    location with good phone and Internet reception.

Chris's story. Chris applied to Microsoft twice, about a year apart in time, before his postdoc and again near the end of his postdoc. The first time he heard nothing; not even a rejection. The second time he got a phone call early the next morning; this turned out to be a complete phone interview. He answered all the questions, and near the end, asked the caller, "So, what is your role in this?" The caller said he was the hiring manager. Surprise! (Eventually Chris was hired, reported to that manager for years, and they are friends today. You never know who might call!)

4. *Does the candidate pass a full interview panel?*
If a phone screen goes well, and the company's needs align with your skills, you'll be invited to a full interview panel. This may require traveling to a company site for 1 or 2 days, although it is becoming more common to conduct these interviews via videoconferencing. We say more about interview panels in Section 12.3.

5. *Can we place the candidate with an opening?* During the interview process, the company and you will each assess the other. On the company side, this is likely to include evaluation of the openings you might fit. It is common for a candidate initially to be considered for one opening, and then switched to another on the basis of interview information. For example, when Chris interviewed at Microsoft, he thought he was being considered for only one position but in fact there were three positions. Ultimately he was offered two positions that fit best and chose between them. It is possible for a candidate to be judged as a "hire" but not to fit any current opening. In that case, the recruiting team may let you know that you will be considered for future openings with an expedited process and fewer interviews.

6. *Does the candidate accept our offer?* Finally, the company will try to make an attractive offer. The offer will have various limitations, such as the available locations and salary range, yet some parts may be negotiable. We discuss negotiation strategies in Section 12.3.4.3.

12.2 Two Formats for Interview Panels

For Quant UXR and similar positions in data science, there are two especially common formats: first, a day-long series of individual interviews, which we will call a *loop* (or *panel*); and, second, a mix of individual interviews plus *hands-on* skill assessment. At the time of writing, the loop format appears to be by far the more popular, although we believe there are good reasons for hands-on assessment as a complement to it.

12.2.1 Format 1: Interview Loops

An interview loop typically consists of three to eight interviews over the course of one or two days. You may expect that half to two-thirds of the interviewers will be in a Quant UXR or closely related role, and their job is to conduct assessment of individual skill areas, aka *hats*. The interview hats cover specific skills such as programming, statistics, and research planning. The remaining members of the panel will represent related *cross-functional* (stakeholder) roles with whom you would typically interact, such as a general UXR, UX designer, PM, or team manager.

Table 12-1 shows an example of how such a loop might progress over the course of a day. The day starts with general orientation, followed by a research presentation (see Section 12.3.2.1), and then one-on-one interviews to assess your skills in core aspects of the position.

Table 12-1. *An Example Interview Loop*

Time	Subject
8:30am	Meet recruiter, orientation
9:00am	Research presentation
10:00am	Research design interview
11:00am	Programming interview
12:00pm	Lunch with another team member
1:00pm	Cross-functional interview
2:00pm	Statistics interview
3:00pm	General interview (often focused on communication)
4:00pm	Wrap up with recruiter

Generally, there is no particular importance to the order of the interviews; it is likely to be determined by the availability of interviewers.

How do the interviewers assess your skills? The organization may have *rubrics*—guides to the levels of skills that they seek—that help interviewers plan questions and assess candidates. Appendix B has a set of recommended rubrics for full-time Quant UXR candidates, reflecting the authors' opinions.

A key point about rubrics and published job requirements is that they reflect a minimum standard, which is not the point at which someone will be hired. Some applicants may be much stronger than the minimum requirement.

12.2.2 Format 2: Hands-On Interviews

An emerging trend for technical interviews in data science, quant UX, and similar roles is to include an extended *hands-on* project in interview loops. You are given a problem and one or more data sets, plus time to perform analyses. At the end of the allotted time, you present your findings to the interview panel, in place of a research presentation.

Table 12-2 illustrates how a hands-on loop might progress over two days. The largest portion of time is allocated to working on the hands-on project. This may be followed with interviews to assess specific skills more deeply, such as reviewing your code or discussing other approaches to analysis.

Table 12-2. *An Example Interview Loop with Hands-On Project*

Time, Day 1	Subject
8:30am	Meet recruiter, orientation
9:00am	Presentation of the project to you
10:00am	Work on the project
12:00pm	Lunch with team members
1:00pm	Check in on the project to answer questions
1:30pm	Work on the project
4:30pm	Check in on the project, discuss initial results

Time, Day 2	Subject
9:00am	Additional analysis
10:00am	Work on the presentation (and final analyses)
12:00pm	Lunch with team members
1:00pm	Finish the presentation
2:00pm	Present results
3:00pm	Additional 1:1 interview, technical
4:00pm	Additional 1:1 interview, general
5:00pm	Wrap up with recruiter

We believe this kind of model is preferable to the interview-only approach because it is a better behavioral sample that assesses your work in a more realistic fashion. It allows you to show what you can do with extended time as opposed to instantly answering questions.

There are several potential variations and choices in this kind of loop. In one variation, you are sent the problem and data in advance, to prepare analyses and a presentation to be given onsite (or perhaps sent in for preliminary review). We *dislike* sending a problem as homework because it becomes an unbounded exercise; this disadvantages candidates who have other demands on their time. It is preferable to interact in person at the beginning of the project to clarify questions and assumptions.

If such a project is given as part of an onsite loop, your recruiter will clarify whether you might use your own laptop (highly advantageous, so you can use your own familiar toolset) or a standard laptop that the company provides.

Although this model is preferable for its realism, it requires substantially greater effort on the part of the organization. Not only must the organization provide the interview problem and data, the interviewers will be required to interact and assess your results in a dynamic fashion. They cannot rely on predetermined interview scripts.

12.2.3 What Happens Among Interviewers?

This is one area where companies vary widely. One model is that each interview is conducted *independently* of the others and interviewers do not converse with one another during the day. Each interviewer gives a rating after their interview, without knowing the ratings given by others. The advantage of this model is that interviewers' ratings are not influenced by others on the panel and are limited to their own observations. A disadvantage is that it may be repetitive, with different interviewers asking you similar or identical questions. The assignment of interviews to specific hats (focus areas) helps to reduce repetition. This model has been common at Google, in the authors' experience.

Another model is a *progressive* interview loop, where interviewers share information and assessment during the day. In this model, the first few interviews might be independent, and then the feedback is compiled to see what areas may warrant further depth in later interviews. The primary advantage of this model is that interviews are able to explore your skills and interests in greater depth, with less repetition. A disadvantage is that earlier interviewers exert influence on those who come later. In the authors' experience, this model was common at Microsoft.

Our recommendation is to ask your recruiter about the model, if you are curious, but overall do not worry about it. In either model, your goals are the same: to demonstrate your skills and knowledge, while learning more about the position.

12.2.4 Who Makes the Hiring Decision?

Some companies allow the *hiring manager* to make a decision (after input from the interview panel), while others send the interview results to a central *hiring committee* that is charged with making a decision on behalf of the company as a whole. Each has advantages and disadvantages. When an HM makes the decision, you will almost certainly meet them and several team members during the interviews, which is valuable. On the other hand, hiring managers may optimize for their own preferences, biases, and immediate staffing needs, rather than considering long-term questions about your potential fit and career. They may also hire a weak candidate if they have immediate needs.

In some cases, another senior manager is asked for a second opinion, adding a whole-company and career perspective while checking the HM. At Amazon, this interviewer is known as the "bar raiser." At Microsoft, this is the "as-ap" or *as appropriate* interviewer. When a candidate passes the interviews with multiple "hire" recommendations, the as-ap interviewer is added to the loop to give a final recommendation.

The hiring committee (HC) model similarly adds a layer of review to examine long-term fit beyond the immediate needs of one group. HC members should *not* have been engaged in the interviews themselves; this allows them to represent a holistic view rather than their own interaction. The HC will consider whether a candidate is *above the bar* (meets the standards) for the company as a whole, and is not merely an expedient choice for one manager.

Two advantages of the HC model are that candidates are considered more completely and the evaluation is less dependent on the recommendation or skill of any single interviewer. On the other hand, it means that you are less likely to meet your potential coworkers—the interviewers may be drawn from across the company—and it adds time to the process. We generally believe that the advantages of this model outweigh the disadvantages, although in practice it can only be implemented at a company large enough to staff both interview panels and a separate HC.

Chris's story. When Chris interviewed at Google—which uses the HC model—he interviewed at the Mountain View, California, headquarters and ultimately accepted a position in New York City. When he arrived on the first day in New York, he discovered that he had not met a single one of his colleagues. Not only did none of his interviewers work in New York, none of them worked on the same product! That reflects a whole-company hiring model at its limit.

12.3 Before, During, and After an Onsite Interview

In this section we discuss details that are relatively small and yet important. They make a difference in your interview experience, perhaps to feel more comfortable, or to help you obtain an offer.

12.3.1 Before: What Happens at the Company

It takes significant effort for a company to do in-person interviews, and only a small part of this will be obvious to you as a candidate. The most obvious part will be interaction with the recruiter to arrange interview dates and online meetings or travel.

The recruiter works with the hiring manager to put together an interview panel. In some cases, especially in smaller organizations, this may be a fixed group of people who are the "hiring team" for one or more openings and interview every candidate. More often, however, an interview panel is formed ad hoc for each candidate.

In the HM model, the panel is likely to include several team members, and will be somewhat consistent across candidates, varying according to team members' schedules and skills relative to each candidate. In the HC model, the panel may be uniquely assembled for each candidate, based on availability of interviewers.

12.3.1.1 Panel Membership

Who will be on the panel? As previously noted, the HM model often includes several members of the team that is hiring. The HC model *might* include team members, especially if the position is in a highly specialized or very large team. However, it also might be drawn from interviewers across the company.

The panel is commonly composed of people with a few different backgrounds, representing key aspects of the company. They are often selected to include a mix of interviewers who are *at* and slightly *above* the candidate's expected job level. For Quant UXRs, a panel might include many or all of the following:

- Two or three Quant UXRs at the same or higher level, who assess technical skill

- A general UXR who assesses user-centered thinking and general UX thinking

- A PM or senior designer who assesses stakeholder interaction and product influence

- A UX manager who assesses general team fit and communication (likely the hiring manager in the HM model)

- In the HM model, another senior manager who "checks" the overall decision

Some organizations do *paired interviews*, where you will meet with two interviewers in a single session. This gives more team members an opportunity to meet with candidates, and in this case, the interviewers will divide the questions between them.

In other situations, especially in larger organizations, the second person may be a *shadow interviewer* who is observing the interview for training, feedback, or calibration (reliability assessment) purposes. In this case, the primary interviewer will ask most of the questions, and will explain this to you at the beginning of the interview. Otherwise, it does not change the interview for you.

12.3.2 Before: Your Preparation

There are several things you'll need to do in advance of an interview loop. We will not recap general advice on interviewing that is available elsewhere, such as handling anxiety. Instead, we have a few notes and advice that are specific to tech companies and Quant UXR interviews.

12.3.2.1 Research Presentations

A research presentation is a common request. It asks you to present a research study to a small group, perhaps five to ten people. The audience is likely to include several of your later interviewers, the hiring manager, and a few UXers, PMs, engineers, or others who might be your stakeholders. Their job is to see whether you communicate clearly and can field reasonable questions about methods, findings, and application of your research.

What makes a good presentation? First, it should be a *research* presentation and not a talk about yourself, hobbies, experience, résumé, education, or past work. Devote a *single* slide to highlight background information of interest and spend the rest of the time on a research project.

We urge you to choose *one* or at most two projects to present. Resist the urge to make it a portfolio of everything you have done or could do. The goal is to demonstrate excellence in one project and the ability to communicate about it.

The presentation should be organized like a talk for a research conference, except less technical. Think about how you might present to a very smart person who is not an expert in your field; that will, in fact, be your likely audience in the Quant UXR role! Start with the motivating problem, then present a brief discussion of the data and methods, and spend the most time discussing the results. When presenting the results, focus primarily on the *impact* of them, if that is known. What changed because of the work? What would it mean for future projects? As for technical detail, present enough to show generally what you did; there is no need to describe or justify every aspect of the analyses. (Candidates from academic backgrounds are especially prone to this pitfall.)

When choosing a topic, pick a project that has *interesting* results rather than technically complex analysis. Again, the goal is to demonstrate communication skill. Engaging the audience is better than attempting to demonstrate technical skill. On the other hand, you might want to have an *appendix* with technical detail, in case someone asks.

The flow of the presentation depends on both the audience and how you open. We like to invite questions and usually say early on, "Please stop me at any time for questions." Some audiences will sit back and observe, and ask questions at the end. Others may jump in after one or two slides and start probing. Don't read much into either style; they probably just reflect the personalities of the interviewers.

12.3.2.2 Requests for Analyses in Advance

Some organizations may ask you to do homework in advance and to come prepared to present an analysis of a problem they give you.

Our opinion is that this is a generally bad idea because candidates vary tremendously in their capacity for homework for reasons unrelated to their skills, including their current job, family, personal life, and so forth. We prefer to do this kind of work in dedicated time during a hands-on interview (see Section 12.2.2).

If you get such a request, and feel OK about doing that work, then we suggest to treat it similarly to a research presentation as noted in Section 12.3.2.1. Start with the question and why it is interesting, then go over the methods, and spend the most time on the findings and what they mean. In this case, you should be more prepared to go deeply into your code or specific analyses if asked. Because the interviewers have prepared a canned problem, they are likely to be interested in those details.

12.3.2.3 Do and Don't

We recommend one thing *not* to do: don't attempt to brush up on technical knowledge such as statistical methods or programming algorithms immediately before an interview panel. Although an interviewer may ask a deeply technical question, your ability to anticipate the exact topic will be low. Even if you do anticipate it, they are likely to follow up with additional questions to probe your understanding, and those would be difficult to answer without real, first-hand experience. In other words, don't treat interviews like preparing for an exam.

How should you spend your time instead? First of all, prepare a good research presentation. Make it as clear and streamlined as possible, and practice giving it. If you can, find someone outside of your field who can be the "smart audience" to judge whether your topic and findings are clear and interesting.

If an interview panel does not require a presentation, you will be expected to discuss examples of past projects in the interviews, and it is worthwhile to practice those explanations. Remind yourself of the non-technical aspects of those projects; for example, any difficult interactions that came up with collaborators and how you dealt with them. Such preparation will be very useful for answering *behavioral* questions, where you are expected to offer examples on-the-spot of situations that demonstrate your interpersonal and leadership skills. These questions often begin with the words "Tell me about a time when..."

After that, we have two other suggestions. One idea is to prepare a *briefing book* on the company where you will interview. What in general can you learn about their products, their customer base, and their current strategy? Do they have "big bets" they are making, or current challenges? Besides educating you about the company, this may help you to understand the context of some interviewers' questions or jargon they use. It will also give you something to look at the night before the interview when you will be nervous.

Closely related to a briefing book is to see what you can find out about UXers and especially Quant UXers at the company. Have they published papers? What are their backgrounds? Although we do not encourage name-dropping during interviews, there may be an opportunity to work in something of interest from this background research. For example, at Google, if a question came up about user metrics that you would assess, mentioning the HEART framework (see Section 7.1) could be part of a good response.

12.3.3 During Interviews

Once again, we don't want to recap general advice about interviewing in person, but we have some tech-specific observations about interview days.

12.3.3.1 A General Approach to Questions

We frequently observe two problems in interviews that are easy to avoid: failure to clarify a question, and excessive confidence in a single answer. In the pressure of an interview, it is common to jump ahead to a solution and miss a crucial assumption. It is also tempting to draw on something you know, setting aside concerns in hopes of projecting confidence or avoiding questions, to an excessive degree.

A simple mnemonic may help you to avoid these problems while displaying the virtues of flexibility and openness. The acronym is *CAMO*: *clarify, assume, multiple options, open*. Let's look at each of those:

- *Clarify*: When an interviewer proposes a complex situation or question, ask a few clarifying questions. These depend on the scenario, and include questions such as: Why are they asking? What is the business reason for the question? How soon is an answer needed? How will the answer be used? Are there resources available, such as colleagues, existing code, or data sets?

- *Assume*: Rather than explaining every detail of an answer, state some assumptions that simplify it. For example, if there is no time specification for a request, you might say something like, "OK, I will assume that I have a week. Here's how I would spend that time." Similarly, for a code problem that involves massive scale, you might say, "I would make it work first on a small subset of the data, knowing that we could scale that up with a distributed approach in the cloud. Here's how I would tackle it for one subset."

- *Multiple options*: Rather than proposing a single best solution, mention more than one approach and how they complement one another to increase confidence in a result. Then say that you will talk about only a single one for simplicity, and answer that part.

- *Open*: If you don't know something, say so. If there are obvious limitations or drawbacks to an approach you've taken, mention them. No candidate is able to know everything and no method is perfect. It is fine to say something like, "I've never done that, and would first want to review the method. However, from what I know, here is how I would approach it." It is also OK to say, "I really don't know. I would consult with an expert there." Of course it is possible that you won't know enough to get the job—and that is also okay. Maybe it is not the right job for you, and it's better to uncover that now. More likely, your honesty and transparency will impress an interviewer and improve your odds.

Unlike *camouflage* aka "camo," this CAMO is not used to hide. Rather, it will help you approach interviews in a professional, thoughtful, and appropriate way.

12.3.3.2 Bring a List of Questions

During the day of the interview, keep a list of questions for the interviewers. Create this in advance and then add and remove questions as the interviews progress. What is it that you would really like to know? One goal for interviews is for *you* to assess the company.

Write down your questions and keep the list with you throughout the day. Good questions will be things that a colleague should be able to answer without particular pressure. What do they like best about their position? What are the biggest challenges? How do they know when they are successful? You can ask these same questions of multiple interviewers, because you may get different answers.

Good questions will make you seem smarter and more interested...and it will rescue you from a situation that many interviewers find uncomfortable, which is when they ask whether a candidate has questions and receive the answer, "No. I don't." Good researchers always have questions.

12.3.4 Afterward

What happens after your in-person interviews? We have already discussed who makes the decision about hiring; see Section 12.2.4 for discussion of hiring manager and hiring committee models. Here we discuss the things that might happen on *your* side after interviews are complete.

12.3.4.1 Thank You

If you like, send a brief but genuine "thank you" message to each interviewer, or at least to ones whose interviews you especially enjoyed. This is not required, nor will it influence their rating of you in any way; it is simply a courtesy. We have been surprised at how few of these messages we receive, yet we always appreciate them.

It can be difficult to know how to contact an interviewer if they did not share contact information with you. Don't be creepy and track them down online! The best idea is to send a note to the recruiter, who will forward it.

12.3.4.2 Fit Calls

If the HM or HC answer is, "Yes, we want to hire you," then you may be invited to a discussion with one or more hiring managers to discuss a specific position. On their side, this may be to answer the question, "OK, the candidate is a fit for the company, but are they a fit for *my* team?" On your side, the question should be, "Is this team a good fit for me?"

Generally these conversations are less technical than interviews and discuss the specifics of what a team is working on and what their needs are. We suggest to be positive while also honest in these discussions. It is fine to ask, "What are you working on?" or "What would success look like in six months?" or even, "Is it a concern that *some method* is not a good fit for my skills?" Transparency in expectations will help everyone. For other questions, see upcoming Section 12.3.4.4.

If a fit call doesn't go well, don't worry too much. Having gotten to that point is a good sign that there may be a better match with another team soon.

12.3.4.3 What You Can Negotiate

A common model for compensation at tech companies is a base salary, an annual bonus, and stock grants (options or outright share grants). Additionally, new hires may receive a one-time signing bonus, stock grant, or a relocation allowance. These amounts will be relative to a *job level* for employees who are comparable in experience. For more about job levels, see Section 14.1.1.

You should first discuss the level with your recruiter (in fact, this should ideally be discussed *before* the interviews). Although we don't like to focus overly on titles and levels, you should feel like you are fairly treated and positioned with regard to your peers.

Once the right level is established, the available salary band might be predetermined, with little room for the recruiter to offer more in terms of salary or stock awards. It doesn't hurt to ask, of course!

However, the amounts offered for signing bonus, one-time stock grants, and relocation may be more negotiable. Those are one-time expenditures that do not affect permanent compensation or ongoing comparison with peers, so a recruiter may have greater flexibility with them, or the ability to reallocate among them (such as lowering a relocation allowance while raising the signing bonus).

Don't expect or demand too much. Although a recruiter may have some flexibility, it may be very limited; you will need to judge that based on how a recruiter responds. Our suggestion: determine in advance what your thresholds would be to say "accept," "decline," or something in-between for negotiation. If you write those down and compare an offer to them, it will make the negotiation simpler.

Finally, we encourage you to think about *time off* before starting. Many positions will want you to start immediately…but that doesn't mean they will say no if you want more time. By the time they have offered a job to you, they are unlikely to walk away simply because you want to wait for a month or two, particularly if you are relocating for the position. And don't worry about whether they will resent it later. After you start, they are unlikely to care about the fact that you started a bit later than you were asked, if they even remember it. We suggest this because time off—with no projects due, no performance reviews on the horizon, and a job lined up—is a rare luxury. When possible, ask for it and enjoy it!

12.3.4.4 Red Flags

You might observe things in an interview process that demonstrate that a position would be unacceptable to you. These would vary from person to person, of course—what one person enjoys might be intolerable to someone else.

We suggest that you *write these down* in advance. That makes it easier to assess during the interview day, and more difficult to dismiss them later. It will also suggest questions to ask of interviewers, when you get a chance.

Some ideas to consider are

- How do you feel about the interviewers as potential co-workers? Do they make you feel respected and welcomed, even if you have a different background from them? How aggressive and challenging are their questions? Do they respond to your questions with openness and honesty, or defensiveness?

- How do interviewers talk about previous colleagues who left the team?

- What is the engagement process with stakeholders like? Ask for examples and more details, such as, "How do stakeholders make decisions using quant findings?" "Do they dismiss negative findings?" "How much Quant UXR time is spent pulling data to answer immediate questions?"

- Does the organization support continued learning, such as conference attendance? How about publishing or speaking?

Again, there is no perfect answer to these; the point is for you to think about what matters to you, in advance of interviewing.

There are a few other warning signs you might observe during interviews, or perhaps in advance of them, as in a job description:

- Do they mention wanting a "rock star"? This suggests excessive expectations. Ask for clarification or examples of what is or is not "rock stardom."

- Do they want a Quant UXR to have expertise across many skills and methods? It is not unusual for job descriptions to take a "kitchen sink" approach and list far more than one person could do. Ask for clarification. The actual expectation may be a subset of the apparent list.

- Do they focus on business metrics (such as revenue or profit) instead of user-centered research (such as behavior, needs, and preference)? There is nothing wrong with attention to such metrics, but they do not align well with most UX organizations (see Section 3.4).

12.3.4.5 When the Answer is "No"

The usual outcome from most applications and interviews is negative. Most candidates don't fit most positions. Even when candidates fit well, there may be too many good applicants, noise in the interview process, or changes on the company side that restrict their ability to make an offer (such as a headcount freeze, which may appear at almost any time due to changes in the economy or a company's strategy).

This means that a negative answer should never be taken personally. Truly, as with the dating cliché, it is more about *them* than about *you*!

When you do well, a company may say that they don't have anything *now* but will contact you again in the future. Sometimes this really happens. Chris once interviewed a great candidate who was not a fit for Chris's current project. Less than a year later, Chris changed projects to one that seemed like a perfect fit for the candidate, who came to mind immediately. He called the candidate again, and she was still interested. Ultimately she joined his team, worked there for several years, and is still a valued colleague—and superstar researcher—today.

As for your performance during interviews, you are unlikely to receive much feedback. Although it is always good to ask, many companies do not provide feedback to candidates as a matter of policy, and might give only very general information such as, "In this case, we're looking for more of ____." Do not read much into signals from any single interview. Perhaps an interviewer grilled you on some technical detail, or asked something you didn't know—despite that, the odds are strong that your response wasn't the ultimate reason for a negative decision. More likely, they were probing the edges of your knowledge and there were other factors that determined the outcome.

Still, what you *should* do is to consider what you liked and disliked during the interviews. Were you put off by deeply technical questions or did you enjoy them? Did the hypothetical research problems seem fascinating or were they outside of your interests? Those reflections will help you assess your fit with future potential jobs, and give you ideas to update your "red flags" list as previously described.

12.4 Job Postings and Applications

As before, we won't repeat general job hunting advice, but we have a few suggestions that apply to Quant UXR and similar positions.

12.4.1 Finding Jobs

Some companies have active recruiters who seek out candidates, and if you are fortunate, they may approach you. However, that happens most often with senior contributors, while other candidates will use typical job search sites, such as Indeed, LinkedIn, and companies' own job boards. When searching, be aware that UX positions have different names from company to company. Potential search phrases include "UX," "user experience," "user research," "usability," "design research," "human factors," "HCI," and "human-computer."

Similarly, the signifier for *quantitative* research positions may differ. Some suggestions are "quantitative," "quant," "statistics," "data science," "experimental," and "scientist." See Sections 3.3 and 3.4 for discussion of different types of UX research positions and how Quant UXR differs from other roles.

12.4.2 Additional Suggestions for Applications

When you put together your application, you will likely be asked for a résumé and references. We have a few suggestions that may be of interest to candidates, especially those who are new to the tech industry.

12.4.2.1 Informational Interviews

If you are fortunate enough to know someone who works in a similar position, ask them whether they will do an *informational interview* with you. This may be a friend-of-a-friend, someone who taught a class, or someone you met at a conference. If you don't yet have personal connections like these, you could also reach out to someone you don't know personally but who wrote a blog post or tutorial that you found particularly helpful. Keep in mind that they may receive many similar messages.

Ideally, their role should be as close as possible to the one you are considering. For example, if you apply to Quant UXR, try to find a Quant UXR or a general UXR, or perhaps a data scientist at the company. Someone in a quite different role, such as software engineer or account manager, would be unlikely to know much that would be helpful.

Your goal in an informational interview is to learn about the daily job, the skills needed to succeed, and whatever the other person can say about its advantages and disadvantages. If they encourage you, you might also ask them for a *referral*, as we discuss next.

12.4.2.2 Referrals and References

Many companies accept *referrals* from current employees. The general idea is that current employees are likely to know excellent candidates from previous jobs, industry events, and school.

To get a referral, ask someone you know in the job. They will want a copy of your résumé, and if you have worked together in the past, they will write a brief evaluation of what they know about your work.

The advantage to you is that this distinguishes your application from one that comes in "cold" from an Internet application. For the employee who refers you, there may be a bonus or other recognition if you are ultimately hired.

We would note several things *not* to do. First, don't ask someone you don't know for a referral (instead, perhaps ask them first for an informational interview, as discussed in Section 12.4.2.1). Second, don't ask more than one employee for a referral; once one person has referred you, you'll be in the system. Third, don't ask for a referral if you've already applied online; you'll already be in the system, and it would waste their time. Instead, if they know you well, list them as a reference. Finally, don't ask the referrer to do a lot of work, such as making multiple referrals for different positions or within a short period of time. Select the positions and referrers carefully, and wait before asking again.

12.4.2.3 Cover Letter

In our experience, cover letters are usually unnecessary for tech jobs and might not be requested as part of the application process or made visible to interviewers.

However, some organizations do take cover letters seriously. If you are asked or have the opportunity to provide one, then do so! It won't hurt and might make a difference, especially if your skills or background are clearer with additional discussion, or you have specific reasons for applying to that organization. Yet if it requires artificial effort—such as prepending a cover letter to your résumé to force it into the system—then skip it.

12.4.2.4 CV vs. Résumé

You'll be asked for a résumé, but you might instead consider submitting a longer CV (*curriculum vitae*, common among academics). The difference is that a résumé is typically one or two pages, while a CV may be substantially longer.

Some companies care a lot about the length, and others don't care much unless the length is excessive. Our suggestion: submit a standard résumé, and then, if you are concerned that it may be too short, ask your recruiter if a longer one would be appropriate or helpful. If it seems OK with them, submit as long of a résumé as you would like, perhaps bordering on the detail that might be in a CV, *if and only if* the additional length is relevant.

For example, if you have publications or patents, it can be helpful to list a few of them, and you shouldn't worry about whether that makes the résumé an additional page longer. On the other hand, there is no need to list every small project, presentation, or job you have had.

In other words, make your résumé as long as necessary to communicate your experience, background, and skills...and no longer.

Do not exaggerate technical skills on your résumé. Anything listed there will be considered for potential in-depth interviewing, especially if it somehow appears to be out of place or unlikely. One of the most common ways to annoy interviewers and hiring committees is to claim expertise that is investigated and found to be lacking during an interview. Unless you are ready to be interviewed about it—without warning, and by any interviewer—leave it off.

At the same time, don't underestimate your expertise when you really do have experience! The goal is not to avoid being interviewed in depth, but rather to be *accurate* about where that should occur. When you express your skills clearly, it will benefit you by steering interviews to the areas of your strengths.

12.4.2.5 Personal Websites and Open Source Projects

If you have a website or public code—such as a GitHub project—expect that interviewers will review it. Don't worry; this is a good thing! Be sure to put such projects on your résumé.

As we discussed in Chapter 6, a key goal in the programming interview is to assess whether you are able to program at all. An open source project provides a signal to interviewers that you do program, and the interview can focus on your interests and the generality of your abilities.

If your project involves collaborators, be prepared to clarify what you did vs. what others did, and set expectations properly. Otherwise, there will be a risk of misunderstanding, such as being asked to use a language you don't know, or to solve a more difficult problem than might otherwise be given to you.

12.5 Key Points

We covered a lot in this chapter, and rather than repeat many details, we will highlight a few things.

Companies vary a lot in their interview practices. You may be able to gain insight up front from informational interviews or, at a minimum, by carefully asking your recruiter about the expectations.

Don't worry too much about the technical details; it is almost impossible to cram for those. Instead, your job is to represent your skills and work well, confidently, and clearly—and to assess the company from your perspective. The best preparations are those that will help you understand more about the role, the company's strategy, and what the actual job might entail.

12.6 Learning More

You are probably wondering, "OK, but what will actually be *asked* during the interviews?" Or, if you're a hiring manager, you might wonder, "What *should* we ask?" Either way, you're in luck. We discuss possible interview topics in more depth in Chapters 4-6. We also list recommended rubrics—an outline of minimal requirements along with more advanced skills—in Appendix B, "Example Quant UX Hiring Rubrics."

As we noted already, we don't recommend trying to learn technical skills in a crash course for interviews, so we don't suggest texts devoted to interview preparation as its own topic. We believe that time is better spent working on real projects, learning more about the field, or improving your knowledge—again, with a hands-on project—in statistics.

Specific recommendations in those areas are

- Hands-on projects with programming and statistics:
  The companion R [25] or Python [127] texts

- Another take on statistics for data science: Bruce et al, *Practical Statistics for Data Scientists* [11]

- For general knowledge of a closely related field: Robinson and Nolis, *Build a Career in Data Science* [114]

CHAPTER 13

Research Processes, Reporting, and Stakeholders

New Quant UXRs will have many questions about working with stakeholders. Who is a stakeholder? How do I engage with them? What should I do when I'm asked to answer a research question?

These questions are especially important for UXRs coming from academia who are unfamiliar with industry positions. It is necessary to meet the expectations of stakeholders in industry, but that does not imply simply doing what they ask. Neither does it mean that you should engage with them deeply about research methods, as you might do with an academic collaborator. You will also need to adapt the way that you communicate findings.

In this chapter we describe common patterns of good and bad relationships with stakeholders. This should help you to form better relationships earlier, be prepared for common misunderstandings, and lead more effective research.

13.1 Initial Engagement

In the broadest sense, a *stakeholder* is anyone who will review or be influenced by research products such as reports and recommendations. For Quant UXRs, typical stakeholders are product managers (PMs), executives, UX designers, software engineers, and fellow researchers. (For definitions of those roles, see Section 2.1.) Put differently, a stakeholder is *someone who wants or needs something from your research.*

© Chris Chapman and Kerry Rodden 2023
C. Chapman and K. Rodden, *Quantitative User Experience Research*,
https://doi.org/10.1007/978-1-4842-9268-6_13

13.1.1 What Stakeholders Want…and What They Need

Stakeholder questions align with the product lifecycle depicted in Figure 2-1, and we compiled some common questions in Table 3-2 (Section 3.2.2). At a high level, four general areas of questions *might be*

- Who are our customers? (Pre-development)

- What do those customers need next? (Early needs)

- Will this product meet those needs? (Development)

- How happy are customers with our product? (Post-development)

We say that those *might be* the questions because the following alternative stakeholder questions are sometimes more common:

- Can we validate that *[some product]* is the right product?

- What do we need to change about *[some product]*?

The difference between the first list and the second list is this: the first list is *user-centered* and asks what customers need, while the second list is *team-centered* and focuses on assessing a predetermined product idea. This follows the reward model in industry: people are rewarded for making and shipping products. Stakeholders want to ship their product as soon as possible, and they hope to "validate" that it is a good idea.

However, as representatives of the business—and as creators for society—stakeholders actually *need* something different. They need honest assessments from users, with as much specificity as possible, of whether *[some product]* will deliver real value in the world. Will it meet an unmet need and add to users' lives?

The team-centered model too often starts from the assumption that a product will meet an unmet need. The team is excited about the product idea and expects that customers will respond to it. However, whereas the team is paid to create the product and is proud to see it, users do not receive those benefits. Instead they must pay for it, in terms of money, time, attention, effort, or emotion. When users do not see value, they will not use or purchase it.

Our job as Quant UXRs is to find a way to help stakeholders understand what they really need to know, even when the system is pressuring them and us to give excessive attention to immediate product assessment.

13.1.2 Focus on Decisions

One way to focus stakeholder attention is to ask about specific *decisions* that need to be made. For a given decision, what actions might be taken? What does the stakeholder need to know to make the decision? What are the risks involved in *not* knowing that information?

Stakeholders often have difficulty specifying decisions that UX research could inform. This is sometimes because their mental model for a UX decision is limited to the concept of A/B testing. Following is an incomplete list of questions—with underlying decisions—that might be addressed through UX research:

- What user needs should we focus on, that have the most value to customers?

- Which problems with our current product should we address in the next version?

- Should we prioritize *this set* of needs or *that set* of user needs? Why?

- With limited resources, should we build *this feature* or *those features*? Why?

- Who should we target as customers for this product? Who should we *not* target?

- Should we build this product at all? Does it meet real user needs?

There are many more questions, and this list is intended to help you broaden your thinking for discussion with stakeholders. We present additional questions and their alignment with research methods in Section 3.2.3.

13.1.3 Work Backward

It is helpful to *work backward* from a hypothetical research deliverable such as an imagined report. This can pin down the point where a stakeholder would make a decision and clarify the data and analyses that are needed for it.

A simple way to do this is to present the stakeholder with possible results and ask what they would do with them. For example, "Suppose customer satisfaction goes down by 2 points. What will we do? What if it goes down by 5 points?" This will answer two important questions for you as the researcher. First, is there a decision at hand, or do you need to clarify that before undertaking research?

Second, if there is a decision, what is the precision that you need to achieve in your findings? In the preceding customer satisfaction (CSat) question, you might learn that a stakeholder wouldn't care about a 2-point change but would care greatly about a 5-point change. That will help you design the study in terms of reliability and sample size.

A more complete form of working backward is to simulate data and build out possible analyses, charts, and final deliverables. That is even better as a prompt for stakeholder review ("How about this report? Would that be useful?") and it provides direct examples to support time or budget requests. However, it takes more work upfront and may be worthwhile on smaller projects only when you know that you can do it quickly or will need the code later. With such a process in place, you may be able to drop in the final data set and achieve almost immediate results or, even better, have time to ponder the results longer before reporting them.

Working backward also forces you to consider how much time you really have to do the research and write it up. If the stakeholder needs to make a decision next week, there is little point in planning to address their question with a research project that will take several weeks. More specifically, our former colleague Michael Margolis suggests asking stakeholders, "What's the latest date that this is still useful to you?"

13.2 Delivering Results

There is no single best or expected format for presenting results from Quant UX research studies. Expectations vary according to the organization, questions, breadth and experience of the audience, and a researcher's own preferences.

We have a few observations and recommendations that may challenge assumptions of some longtime practitioners. These may be useful for academic researchers and others transitioning into industry research.

13.2.1 Stakeholders Are the Users of Your Research

It is useful to think about the outputs from your research as a designed *product*, and your stakeholders as the *users* of that product. We have already mentioned that what stakeholders want from you is not necessarily the same as what they need, which is consistent with the analogy between products and users.

Depending on your background, but especially if you come from academia, you may tend toward authoritativeness and detail in your written reports. Stakeholders and other team members may give you positive feedback, but it's also important to observe whether they take action based on your findings. If they do not, your research is not having the impact that it could be, and a likely reason is that it is not taking stakeholder *needs* into account.

So how do you know what they need from your research? You can use your skills as a researcher to figure this out! Treat your work output like a product, and try to understand the context in which your "users" will experience your "product." This will help you develop a sense of empathy for them, based on that context, which will make it easier for you to adapt your work according to what they need from you. For example, if you are targeting someone who is an executive, they may be spending their day going from one meeting to another, trying to come up to speed quickly on every topic that's put in front of them, and then trying to make good decisions with limited information. They probably see hundreds of emails or Slack messages every day, perhaps while also trying to keep up with meetings, or on their phone in between meetings.

Consider how to adapt the way you present your work, according to that context. Does it seem appropriate to deliver a long and detailed report to the executive in the preceding example? If your team is seeking approval from that executive for a specific product decision, understanding their needs should help you realize that you have to do more than simply walk through your research methodology in detail. Help them feel confident in making the decision by focusing specifically on how your research findings inform it.

Going beyond that, consider how easy your work products are to reuse and reference, even when you are not there to explain them, or to remind people of the link. If a PM has to make a case to executives, based on your work, how well are you supporting them in doing that? How easily do each of your slides or charts stand alone, to be easily pulled into a presentation in a different context? Could you make your findings more memorable, such as with a diagram for easy recall and sketching on a whiteboard? For example, thinking of a suitable acronym helped make the HEART framework (Section 7.1) more memorable and, consequently, more useful.

13.2.2 Two Models: Presentations and Documents

There are two common models for presenting research results: a *presentation* deck created with Microsoft PowerPoint, Google Slides, or similar software; and a written *document* created in Microsoft Word, Google Docs, or another text editor. We'll look at the advantages and disadvantages of each model.

A third possibility is an engineered product such as a dashboard or another kind of interactive, updated visualization. In this section, we discuss static deliverables; see Sections 7.2.3 and 8.2.7 for discussion of dashboards.

13.2.2.1 Presentation Slide Decks

A *slide deck* is the most common form for presenting research results in industry. When we were working at Google (and at several other places), this typically worked as follows:

1. The researcher puts together a presentation with the background and key results soon after analysis is complete.

2. This is reviewed informally by a few partners—perhaps fellow researchers on the team, designers, the PM partner, or one's manager—to improve it.

3. There are two or more *shareout* meetings: one to the immediate, smaller team of stakeholders, and another to a larger group of everyone interested. This may be recorded for others to view asynchronously.

4. The presentation is shared concurrently via email, Slack, or some other method that the organization prefers.

5. Questions and answers are addressed live during the presentation, through comments on the slide deck, and via email or Slack.

6. Additional materials such as code are linked in an associated document, notebook, or other format as relevant.

7. The final presentation becomes the canonical archival document for reference.

Advantages of Slides

There are two primary advantages of the presentation model. First, the slide format allows great flexibility in the materials and visual style. The researcher can easily include charts, videos, recordings, overlays, annotations, callouts, links, and other elements that highlight the elements studied and emphasize important results. Second, the linear nature of a live presentation gives tremendous control to the researcher to shape the narrative and audience experience. A good presentation is an engaging and memorable way to convey findings.

Disadvantages of Slides

There are significant downsides to the presentation model. One is that it is difficult to tell a story in archival slides. Too often slide decks are confusing, difficult to read, and dependent on live performance. When they are optimized for archival reading, they become laden with text that leads to boring presentations.

Another problem is that senior stakeholders are likely to jump in and pepper a presenter with questions after only one or two slides. If they have a link to the presentation, they may skip ahead and start reading the results slides while the presenter is still talking about the background. The linear narrative is immediately derailed, and it can be difficult to regain control of the presentation order.

Finally, the presentation model depends on the live performance skills of the researcher. Someone who is confident, experienced, or extroverted may give a convincing presentation even if the underlying research is of low quality.

13.2.2.2 Research Report Documents

An alternative to slides, which Chris has used at Amazon, is to share research findings in a written *document* similar to a research article or white paper. Such reports may be authored in an editor such as Microsoft Word or Google Docs, or by using a notebook or markdown system such as R Markdown or Google Colaboratory.

This approach has long been used at Amazon. The general steps are exactly the same as in the steps noted for slide presentations in the preceding section, with two exceptions:

- The deliverable is a written document, not slides, and it is limited to no more than four to six pages. It should contain only the important points and findings, not technical details.

- At a presentation meeting, time is reserved for the team to *read* the document in its entirety. For example, the first 15 minutes of the meeting may be reserved for reading in silence. This is followed by questions and discussion. No slide presentation is necessary.

Advantages of Documents

The document format (with reserved time for reading) has two primary advantages over the presentation format: it emphasizes clarity, brevity, and actionability of the reports; and it ensures that a team reads the material before jumping in with questions that disrupt discussion. This style also deemphasizes presentation skill in favor of writing skill, and ensures that archival documents are readable without requiring personal narration. A well-written document can convey information more efficiently than a presentation.

Disadvantages of Documents

The main limitation of written documents as research deliverables is their inflexibility in format. Depending on the platform, it may be difficult or impossible to include video or sound clips, animations, very detailed images, annotations on charts, and the like, all of which are easy to include in slide presentations. The severity of this limitation depends on the nature of the findings to be conveyed. For those reviewing a document outside of the original meeting context, it may be harder to skim than a presentation. Also, too many documents are poorly written and unclear (although the same is true of slide presentations).

13.2.2.3 Recommendation for Reporting

Overall we find that the advantages of document-based reports outweigh those of slides, provided that you reserve time during the meeting to make sure everyone has read the document. Documents encourage brevity and clarity, and they help ensure that an audience engages with the *findings* more than the *presenter*.

If your organization uses slide decks, we encourage you to try documents as an alternative. This will be a cultural change and the following points may help:

- Provide templates and examples to demonstrate what you're looking for.

- Have managers review and give feedback on documents before they are shared.

- It is essential to set aside dedicated time in meetings for silent review of the documents: at least 10 minutes and possibly 15 or 20.

- Provide an appropriate system for collaboration so that versions of documents are not circulated as email attachments. Google Docs provides outstanding collaboration mechanisms.

For examples of reports and reporting templates, check the resources at this book's website, `https://quantuxbook.com`.

13.3 Principles of Good Deliverables

Whether your research reports are delivered as presentations or documents, several principles will increase your research influence and make stakeholders happier.

13.3.1 Short and Focused on Action

The first principle is to focus relentlessly on what your audience needs to know to take action. Make your report or presentation short and keep it directly oriented to information that stakeholders need to make a decision.

An effective way to do this is to outline the key questions you will answer at the onset of a research project. Review them with stakeholders before beginning data collection (when possible) and use them to guide your research along the way. Then list the questions at the beginning of your report, followed by short answers.

An *executive summary* is helpful and much appreciated. This summarizes your key research findings in a paragraph or two, or on a single slide, and is similar to the abstract of an academic paper. Does it worry you that the audience will find their answer and skip important details and narrative path of the complete document? It shouldn't! More likely, a summary will help the audience know what to expect and to process the details more effectively. The summary should deliver the most important messages succinctly and ensure that the audience reads them.

These principles also apply to the charts in your reports or presentations. It is common, especially when under time pressure, to think that it is not worthwhile to prioritize chart design, such as adding annotations or highlighting key data points. Elements like these often require detailed customizations, which can be time-consuming to create, and it is tempting to simply reuse whichever default charts you created during exploratory data analysis. However, the insights *you* gained from such charts may not be at all obvious to a stakeholder. Careful consideration of your stakeholders' needs (see Section 13.2.1) and thoughtful design choices (such as choosing a simple chart type, using text and color to emphasize key findings, and removing unnecessary clutter) are extremely helpful in ensuring that your charts actually convey your findings in a way that leads them to have impact.

13.3.2 Minimally Technical Reports

In industry, a research report is not a place to demonstrate technical excellence, complexity, or achievement. Nearly everyone in your audience will be less proficient at the methods than you are, will not understand the complexities of the data, and will not know when to be impressed or unimpressed by what you did. What will impress them are clear and actionable results.

This means that your report should contain *none* or almost none of the following:

- Details about statistical models

- Equations

- Checks on convergence, quality, or reliability of findings

- Output from statistical models such as coefficient summaries from regression models

- Interpretations of statistical significance

- Plots that assume statistical knowledge such as box-and-whiskers plots, QQ plots, leverage plots, correlation matrix plots, scatter plot matrices, and so forth

What should you show instead? We suggest the following:

- The key questions and answers.

- *Brief* summaries of the data sources and collection and analysis methods. We emphasize *brief* because details should go elsewhere, as noted following this list.

- Plots that reinforce the key answers (and only the key answers).

- A sense of whether there is high or low uncertainty in your results. (This is often assisted by plots with confidence intervals, but do not dwell on interpretation of what "significance" means.)

What about technical details? Are they not important? Yes, they are important—so *link* to them for anyone interested. Include a brief technical appendix in your report with a paragraph or two that gives an overview of what you did. That appendix can point to your code, data sets, or other technical details. That way an interested reader can find the details when they wish. In our experience that is usually only one or two people for any given project. Everyone else will be reassured to know the details are in place and also relieved to skip them.

13.3.3 Remain Unbiased

Your job as a UX researcher is to help a team listen to users and improve products and services. Your job is *not* to promote a product, to help it succeed independently of user needs, to make your team happy, or to prove your own value. Those outcomes are valuable but they should occur downstream from focusing on user needs and communicating them to the team (see Section 4.2).

Your reports should focus carefully on communicating what you actually believe you have learned from users through your data and analyses. This often goes awry in two ways. First, too many researchers are hesitant to communicate what they really believe and use vague words to couch everything in statistician speak. This leads to statements such as, "The data in this sample show a trend that is consistent with many users showing interest in X, p < 0.10."

Don't do that! Instead—if you believe it—simply say, "Users want X."

Second, researchers are often afraid to deliver the bad news that users don't like a design, don't want a product, won't pay for it, see no need for it, or aren't using it in the way the team hoped. It is easy to suggest improvements instead of delivering tough news. But if you believe that the bad news is the real news, then deliver it. (We say more about this from a research ethics perspective in Section 4.5.5.)

13.3.4 Reproducible and Generalizable

Your analyses should be *reproducible* and *generalizable*. Those terms have a narrow meaning and a broad meaning. The narrow sense of *reproducible* is that any other researcher, given a different sample from the same data sources and using the same methods, would obtain consistent results. Even more narrowly, it means that using your data set and code, the results would be identical. A typical threat to reproducibility is performing actions in an accidental sequence, such as setting a temporary data filter and forgetting to undo it. If you script your analyses and routinely run them from the top, they will be repeatable.

The broad sense of reproducibility and generalizability involves your assessment of whether the high-level answer would be expected again, if a study were repeated with different samples, methods, or conditions. In other words, do you really believe the answer? Do you expect that another researcher, using different methods, would arrive at a similar answer for the business?

If you believe that, then communicate it clearly in your report. Or if not, say why not and state the limitations or recommendations for additional research. *You are the research expert* and your organization expects you to determine the importance and reliability of findings.

13.4 Research Archives

A research archive is a collection of presentations, documents, and supporting material that details UX research for a team or organization. The goals for a research archive include collecting work in one place for easy access, finding previous studies of a particular product or topic, demonstrating the work a team is doing, building a sense of knowledge accumulated over time, and providing examples for new employees and projects.

Some of those goals are more difficult to achieve than others. In particular, it is surprisingly difficult to find work related to a product or topic in a research archive, for several reasons. One reason is the problem of *base rate*: there are so many possible topics and products that even an archive with tens of thousands of research documents will have little or no coverage of a random topic. Another issue is that people use different terms, even for the same product, and especially for concepts. Many topics are vague, and most research reports are of historical rather than future interest.

To address those issues, an archive's creators may design a tagging system to classify reports. However, maintaining such tags is not only subjective but labor intensive and unrewarding. Researchers will soon stop maintaining them.

If your organization is serious about learning from the knowledge it has accumulated over time, we recommend that it should hire full-time, dedicated curators (such as a *UX librarian* or *knowledge manager*) to design and maintain a research archive. Otherwise, a centralized archive may prove difficult to use in practice.

Instead, we suggest that individual researchers maintain a personal, curated, and lightweight research archive document. This personal archive should categorize the researcher's most important work with brief summaries and links to complete documents. Table 13-1 shows an example with the crucial information: the topic, data, brief summary, and link.

Table 13-1. *Personal Research Archive*

Topic	Date	Overview	Full Report
CSat	10/2022	2022 Quarter 3 CSat with driver and segment analysis	//link/report1
CSat	04/2022	2022 Quarter 1 CSat	//link/report2
User Needs	06/2022	In-home evaluation of family calendar usage	//link/report3
Pricing	08/2022	Conjoint analysis of pricing for calendar subscription	//link/report4

A personal archive of this sort is useful for many purposes: sending to team members who want to find research of interest; making the documents easy for you to locate in the future; and keeping a running list that will serve as a reminder for performance reviews. Individual lists may be aggregated to represent the work of a team or larger research organization.

The most important requirement for such an effort is this: it must deliver enough value to the creator—an individual researcher—to offset the effort to maintain it. A simple, brief, curated document as shown in Table 13-1 satisfies that requirement better than a centralized database or complex archive.

13.5 Common Problems with Stakeholders

There are problems with stakeholders we see repeatedly in organizations where we work, consult, and mentor other researchers. In this section, we share suggestions on how to minimize or avoid them.

13.5.1 Lack of a Decision Criterion

As we've mentioned many times already, research is most effective when there are clear questions with specific decisions to be made. A common problem is when a stakeholder requests a project "to see what the data say."

Here's an example. Imagine you're designing eyeglasses and could make them in different weights, where lighter frames would cost more to produce. It would be reasonable to learn how strongly users prefer lightweight frames. A stakeholder might ask you to collect data on how strongly users endorse needs such as *I want lightweight eyeglass frames*.

But what would you do with that data? Will it answer any important question? Or will it not say that, for many users, lighter is better? You know that already.

A better approach is to pin down the exact question being asked. For example, suppose the business needs to determine the appropriate amount it might be able to charge, which will influence the choice of materials. In that case, the question may involve users' willingness to pay relative to weight (perhaps a conjoint analysis study). Or you might need to understand the ergonomic impact of the eyeglasses and determine when a frame is too heavy or too light (perhaps through a field trial with prototypes). Unless you know the real question, your findings ("lighter is better") will not be useful to the stakeholder. Stakeholders are not always good at specifying such questions, and it is your job to help them.

13.5.2 Ad Hoc Projects

Some stakeholders view UX research as a service to be ordered on demand rather than planned in advance. For instance, they might realize during a weekly leadership meeting that they do not have enough information to decide between two design options, and call on UX research to deliver a recommendation before the following week's meeting. In a more extreme model, they may view Quant UXRs as providing a data-on-call service to go and "look up" answers immediately.

There are many problems with this situation. Short-term research of this type is very costly in time (and money if it involves external vendors or suppliers). It is disruptive to the team, and is dismissive of the skills of senior researchers. And paradoxically it often leads to more questions than it answers because the research is designed without forethought, pre-testing, or coordination with other projects.

It is better to plan research on longer time horizons, giving a team opportunities to combine projects for efficiency, and to cover more topics than short-term A/B evaluations. Teams can then design a process in advance to iterate on shorter timelines in the appropriate stages of the product lifecycle.

13.5.3 Opportunity Cost

Of course, even with the best planning processes, unforeseen high-priority questions will still arise, and you will have to decide whether or not to tackle them. This includes asking stakeholders to consider the *opportunity cost*: if a research team takes on immediate requests A and B, what are the projects C, D, E, and F that they will not do?

Again, as with the importance of establishing longer research plans to avoid ad hoc projects, a researcher or research manager should always bring up the question of trade-offs among projects and the opportunity cost of committing to any one of them.

13.5.4 Validation Research

In some organizations, UX research may be requested in terms of "validating the product direction" or "validating the design." As we discussed in Section 13.1.1 this occurs because of a stakeholder focus that is more team-centered than user-centered.

In this situation, we would work with the team to focus on a more specific (and user-centered) question that they wish to answer. Perhaps they need to know whether the design is more understandable than an alternative; or whether it is ready to ship to users; or whether it will meet the business goals for adoption.

13.5.5 Statistical Significance

"Statistical significance" does not mean what most stakeholders believe it means. They commonly interpret it to mean *important,* or *true,* or *repeatable.* Even if they understand that it relates to expectation under resampling, they are unlikely to consider the many other sources of uncertainty. We say more about this in Section 5.3.3.

Our recommendation is to avoid mentioning or explaining statistical significance. In particular, do not use the term or flag it on charts, reports, and so on. Instead, focus on real-world effects. Present confidence intervals wherever they may help stakeholders to focus on the important findings in the data.

Here's an example. Suppose CSat (see Chapter 8, "Customer Satisfaction Surveys") decreases from 81 to 80 points in a sample of N=200. It is natural for a stakeholder to ask "Why?" and, among the more sophisticated, "Is that statistically significant?"

Although the answers to those questions are "It's probably noise" and "No, it's not statistically significant," it is better to avoid the whole issue. We might have reported an initial headline answer that there was "no change," and then, in the details, note that CSat has gone from "81 (CI=75–86)"—using a 95% confidence interval for the score— to "80 (CI=74–85)." (Note that this assumes a simple time-over-time comparison. See Chapter 8 for additional discussion.)

13.5.6 Cherry Picking Results

When you share findings, worry a *lot* about the tendency of readers to select facts and figures that reinforce their own views, preferences, or agenda. That doesn't mean they are nefarious; it means they are human.

There is no perfect way to combat this, but we find that three things help:

- Clarify the questions and decisions up front, so the findings can focus squarely on them.

- Be extremely clear in your presentation and unambiguous about recommendations.

- Remove technical detail and supporting data that are not directly focused on the research questions.

A *worst* practice here, which we unfortunately often observe, is to compile and present dozens of tables that slice data in many different ways. For example, a report may detail multiple UX metrics such as adoption, retention, satisfaction, and recommendation likelihood; compile each of those across multiple products or features; and then further break out each combination by several additional factors such as demographic categories, user segment, and business metrics.

Such a report violates all three of our suggestions: it has an unclear focus; it does not demonstrate the specific actions anyone should take; and it presents excessive amounts of information. The most likely outcome is that stakeholders will simply choose the findings that are of interest to them. And who could blame them? Better reporting will help them to stay focused.

13.5.7 Conflicting Results

When there are multiple studies in an area, it may occur that different methods, samples, data sources, or analyses will give conflicting answers. This happens for many reasons, but a particularly common situation is when users' *stated* behaviors or preferences conflict with their *revealed* (actual) behaviors or preferences. For instance, users may report product usage on a survey that is wildly inaccurate with regard to observed usage data in product logs. Or users may express a high willingness to pay for a product that is later contradicted by market evidence or by a more specific method such as conjoint analysis.

Stakeholders' expectations are often informed by experience with physical sciences and engineering. They may expect that measurement of human attitudes and behaviors will be accurate in ways that are similar to physical systems. In such systems there may be an underlying "true" value that can be measured with high conformity across samples or methods.

It is important to prepare your audiences to understand that human behavior is quite different from physical systems and that varying results usually do not *conflict* as much as they highlight different aspects of behavior. For instance, stated willingness to pay for a product may be high because users truly do perceive high value. And yet revealed value may be much lower because actual purchase behavior is constrained by factors such as budget, availability, and competition with other goods (both direct product competitors and the *outside good* of using time and money elsewhere).

We don't recommend attempting to educate stakeholders about the nuances of social science data and methods. Rather, we attempt to set appropriate expectations, highlight when data involve perception or attitudes as opposed to actual behavior, and explain discrepancies up front. It is often possible to present a larger, more complete story that reconciles the apparent differences while providing additional insight into users' needs.

13.5.8 Challenge Only If Negative (COIN)

A final problem is that stakeholders give asymmetric attention to results they don't like or didn't anticipate. If a stakeholder wants to make X, and research says, "Users like X," the stakeholder will usually accept the results gladly. But if the research says, "Users don't like X," then stakeholders may challenge the study design, the sample, the questions asked, the statistical analysis methods, and details of the presentation, and call up contrary evidence (such as conflicting results, described in the previous section). With thanks for the name to Chris's former PM colleague, Eric Bahna, we call this situation *challenge only if negative* (COIN).

There are two general prerequisites for COIN. First, it occurs when stakeholders are not bought into a research plan or a researcher's credibility. This may occur because a stakeholder decides to challenge a plan post hoc, or because it was infeasible to consult them in advance. This often happens in large organizations where research is widely disseminated and yet widespread consultation is impossible due to the large number of interested parties. Second, COIN occurs when someone dislikes a result.

COIN presents both a methodological problem and an organizational problem. Methodologically, it introduces directional bias into scrutiny of results; scientifically one should give equal scrutiny regardless of the direction of findings. In fact, it might be smart to pay *more* attention and scrutiny when results align with your own preferences; this would help offset the known bias introduced by COIN.

From an organizational perspective, COIN signals to researchers that they should not present negative findings and that their work will be challenged if they do. That biases the kinds of research that are undertaken and creates conscious or unconscious motivation to deliver only positive results (see Section 4.5.5).

To combat COIN we recommend that researchers prepare for it in advance. At the onset of research, plan to use multiple methods for important questions. If the results are negative, you can point to the more complete set of findings that uses multiple data sets and methods. Organizationally, we recommend to avoid taking on projects that are likely to run into this problem, either because they involve obviously poor products or features, or because they involve a stakeholder prone to COIN. (On the other hand, if a product may be detrimental to users, it may be ethically important to work on it and advocate for users; see Section 4.5.)

In the case of negative results, get support from colleagues and your immediate management to ensure that you do not become the sole target for COIN. More broadly, as we mentioned in Section 7.5.2.1, this issue is a sign of a broader organizational culture that is unwilling to openly acknowledge failures and learn from them.

13.6 Finding a Great Stakeholder

Despite the challenges we described in the previous section, one of the most rewarding parts of being a UX researcher is working with great stakeholders who value your work and wish to learn from users. We have a simple recommendation: keep working with them.

You may find one of these partners randomly, perhaps a PM who responds well to research and brings you great questions at the right time. In other situations, it may take some effort. One tactic is to interview potential stakeholders. What would they like to know from research? How would they use the answers? When you find someone whose questions and approach align with your own, do what you can to work with them.

When you find such a partner, don't hesitate to propose higher level, strategic research questions or ask for their own ideas about this. They want to tackle larger projects just as you do, and there is a strong likelihood that your research can help them while they also help you in return (with both high-impact projects and support for performance reviews).

Over the course of our careers it has been our great fortune to have had several of these relationships, and they have been professionally rewarding as well as enjoyable on a day-to-day basis.

13.7 Key Points

Stakeholder engagement is the most important part of any role as a researcher, but you cannot expect that stakeholders will be technical experts in research. That is your job, while their job is to be expert in design, engineering, product management, people management, or strategy.

A few things to remember are:

- The questions that stakeholders ask are not necessarily the right questions for research. You must help them to shape the questions (Section 13.1.1).

- An excellent way to frame research is to engage with specific decisions to be made (Section 13.1.2).

- It is helpful to work backward from a research deliverable. Show a stakeholder a preview of what the results might be—using simulated or preliminary data—and ask whether it is helpful before investing in the full project (Section 13.1.3).

- Think of your research outputs as a product, and your stakeholders as the users of that product. Design for their needs, accordingly.

- Research reports are most commonly given as slide decks (Section 13.2.2.1), but there are significant advantages to using written documents instead (Section 13.2.2.2).

- Documents emphasize findings rather than presentation skills and reduce the number of randomly sequenced questions. However, researchers cannot change an organization's decks vs. documents culture on their own.

- Good reports are short, focused on action, minimally technical, and unbiased (Section 13.3). A particular danger is introducing bias in the hopes of making research more palatable.

- Research archives are a great way to compile, reference, and share past research reports. However, too often they are overly complex and cumbersome to maintain. A simple alternative that is valuable to the researcher—and therefore more likely to be maintained—is a curated "one page" list of projects (Section 13.4).

- There are several ways in which stakeholder questions go awry, usually because stakeholders are unclear about the exact nature of UX research. These include ad hoc questions, lack of decision criteria, and challenging results only when they are negative. These are generally improved through focus and engagement early in the research cycle (Section 13.5).

13.8 Learning More

Every organization is different, and you will learn to work with stakeholders over time. Still, the following activities will help accelerate your understanding:

- Ask whether there is a research archive for your team and read through reports from various projects.

- Attend colleagues' research presentations, even if they are not directly related to your work, and observe the interactions with stakeholders.

- Make this a topic for mentoring discussion with colleagues; ask about typical factors for success or problems with stakeholders.

- Ask colleague Quant UXRs or general UXRs to provide feedback on your reports before sharing them. When possible, seek out colleagues who have been on the team longer than you have or who know your audience well.

- Find one or two stakeholders who are especially close to your work— perhaps a PM or designer—and give a preview presentation to them before any larger presentation or sharing. Besides gaining feedback, it will avoid surprising your closest colleagues.

For additional learning, we have a few recommendations. Christian Crumlish's book, *Product Management for UX People* [37], is highly recommended for its insight on how UX relates to the broader needs, goals, and work of product development teams. If you are new to a UX role in a tech company, you should read it.

You could also learn more about the challenges that stakeholders face in their own decision-making. This can help you understand them and suggest potential directions for new research. Kaplan's *The 360° Corporation* [65] investigates the trade-offs that stakeholders must make among business results, employee demands, social betterment, ecological responsibility, regulatory requirements, and other outcomes. Reading it will broaden your view of what business "success" is, and should inspire ideas about how UX research can contribute beyond assessing products.

We emphasized that technical writing for stakeholders should be simple, clear, pointed, and as non-technical as possible. Greene's *Writing Science in Plain English* [53] is an excellent primer and refresher on writing well about research.

For more on conveying findings effectively with charts, we recommend *Storytelling with Data* by Cole Nussbaumer Knaflic [71]. It has many examples that demonstrate why it is worthwhile to put time and effort into chart design and description. Despite its particular focus on Excel charts, the principles can be applied to data graphics generated via any tool.

Finally, the annual Quant UX Conference (Quant UX Con) is an unparalleled, unique conference for Quant UXRs. It features presentations from many Quant UXRs and their colleagues, across a wide array of organizations, discussing industry research problems. Many talks also address stakeholder engagement and career development. For information on Quant UX Con, see `https://quantuxcon.org`.

CHAPTER 14

Career Development for Quant UX Researchers

Quantitative UX research roles have existed for less than two decades, and longer-term career trajectories are only now beginning to emerge, making it harder for us to give targeted guidance in this chapter than in the rest of the book. We will look at typical tech industry models for both general and quantitative UX researchers. We share observations that are specific to Quant UXRs wherever possible.

Much of this chapter will be familiar to readers already working in industry. We take time to share concepts, terminology, and expectations for newcomers. However, we hope that our observations on common problems and career growth will be useful even to senior industry researchers. The discussion is oriented primarily to individual contributors, although it should also be of interest to managers who wish to help Quant UXRs grow.

This chapter contains more personal opinion and observation than other chapters in this book. Our advice and descriptions are intended to be informative for the most common situations we observe, yet they cannot apply to every individual case. Use your own judgment when deciding whether our advice is helpful.

Our overall message is this: Quant UX is an exciting and dynamic area, with many possible paths rather than a well-worn road. There is no definitive guide to a career because we are all defining that together. Your own interests, enthusiasm, and accomplishments will shape the role for yourself and all Quant UXRs.

C. Chapman and K. Rodden, *Quantitative User Experience Research*,
https://doi.org/10.1007/978-1-4842-9268-6_14

14.1 Elements of Career Paths in Industry

In tech companies with UX researchers, there are generally four formal and documented elements that are important in a career:

- *The hiring standards and process*: We discuss hiring standards in Chapter 12, "Interviews and Job Postings," and there is a hypothetical set of rubrics in Appendix B.

- *The tech levels*: These define the job level, title, pay range, and general scope of responsibility. For instance, there could be levels for entry level vs. senior employees that apply to all roles. We outline these in Section 14.1.1.

- *A career ladder*: This defines the requirements for a specific role, relative to the levels. For example, it would define the differences between a UX researcher (a possible entry-level role) vs. senior UX researcher. We describe career ladders in Section 14.1.2.

- *A track within the career ladder*: The tracks encompass variations that are possible as a UXR moves up the career ladder. Most commonly there is a difference between *individual contributor* (IC) tracks and management tracks. We discuss IC and management tracks in Sections 14.1.3 and 14.1.5.

The details of the levels, ladders, and tracks vary from organization to organization. However, there are commonalities that we examine in the following sections. Don't worry too much about the exact titles, levels, or other specifics. Those vary while the overall *structure* is quite similar across many organizations.

14.1.1 Job Levels

There are typically about six to eight levels that encompass most employees in UX organizations and they are typically numbered, with associated job titles. When we worked at Google, the individual contributor levels for UX researchers were those shown in Table 14-1.

Table 14-1. *User Research Levels at Google (as of Chris's time in 2022)*

Level	Common Job Title
L2	UX Researcher I / UX Research Assistant
L3	[Quantitative] UX Researcher II
L4	[Quantitative] UX Researcher III
L5	Senior [Quantitative] UX Researcher
L6	Staff [Quantitative] UX Researcher
L7	Senior Staff [Quantitative] UX Researcher
L8	Principal [Quantitative] UX Researcher

In Table 14-1 we put "Quantitative" in brackets because it may not be defined separately from the standard UXR levels. Quant UXRs may be *hired* according to specific standards for Quants, but after that Quants are leveled in a standard way alongside UXRs of all kinds.

There are various other schemes across the industry. For example, at Microsoft there are ten levels (58–67) that span approximately the same range as five levels (L3–L7) at Google. Levels are ordinal and vary in range from one job category to another. (This is why there is no L1 for UXRs at Google.)

Someone's exact job title is not necessarily defined by the level. For example, an L4 UX Researcher might have any of these titles: "UX Researcher," "Quantitative UX Researcher," "Ethnographer," "Survey Scientist," "Research Scientist," among others. In some companies, the job title is largely a matter of personal preference, and employees can choose how they are described on their business cards or in the internal directory. One of our superb Quant UXR colleagues at Google called herself a "Data Exhaust Analyst."

Titles also do not align precisely across companies; a "principal" or "director" at one company might be a "staff researcher" at another. The website https://www.levels.fyi has user-contributed information about levels at tech companies (caveat: our experience is that it is highly imprecise as to salary, but useful in terms of relative, directional comparisons).

14.1.1.1 Levels, Responsibility, and Expertise

Each level has a somewhat nebulously defined *scope* of responsibility. For instance, when we were at Google, UXR L4 was generally associated with research related to specific product features, immediate product releases, and shorter term projects (such as 1–3 month projects). UXR L5 was associated with projects that might inform multiple features or an entire product in the medium term (3–12 months). UXR L6 was expected to demonstrate work that informed an entire product and also showed leadership or expertise at a cross-company level.

Similarly, UXRs were expected to show a broader range of skills at higher levels. UXR L4 was expected to demonstrate deep expertise in a few research methods (like survey research or A/B testing), while UXR L5 was expected to be strong in a larger set of methods. This might be phrased in terms of "ownership," as in this statement: "L4 could own research for feature testing, while L5 could own the quant research agenda for an entire product."

A detrimental and pernicious, yet common, misuse of levels is when employees gauge the importance or seriousness of a co-worker's research on the basis of level. It is unfortunately common to hear comments like "He's an L7" as if that made his research more important, or "She's only an L3." *Never do that.* If someone else does, find a way to politely challenge them on it.

14.1.1.2 Levels and Compensation

Levels are associated with compensation *bands* that define the range of possible base salary, bonuses, stock awards, and so forth. This is one source of a problem we discuss later, the belief that "a higher level is always better" (Section 14.2).

We have two observations here. First, although you might never be told this, the compensation bands are typically available on request from HR. We won't go into the legal aspects of that, which vary substantially from state to state and country to country. However, we would say that HR is accustomed to being asked about compensation and they won't think less of you for asking about your salary range.

The second observation is that the levels are often closer together than you might expect. Chris once received a promotion from one level to another with only a 1.6% increase in base salary. (More common is a 10–20% increase.) As we describe later in this chapter, once your compensation is enough to meet your needs, we believe you should concern yourself more with your happiness in the work you're doing than with promotion for its own sake.

14.1.2 Career Ladder

A *career ladder* defines job duties, expectations, and criteria for each level in a job role. This may be supplemented by examples, information on promotion requirements, materials for learning or reference, and so forth.

If you hope that a career ladder will tell you what to do as a Quant UXR, you will be disappointed. Ladders are typically vague and stuffed with generic phrases such as "designs and conducts research that is appropriate to the product need" or "produces timely and actionable research summaries to impact design and engineering decisions." (We made those up, but they are illustrative.)

Because career ladders are vague, we don't have a lot to say about them except that they are valuable when it's time for performance reviews and promotion (see Section 14.3). They are also useful for *midpoint* check-ins with a manager in advance of performance reviews, when you should discuss whether you are meeting your team's expectations. For performance reviews and promotion, you will want to cite the career ladder explicitly and line up your accomplishments against it. Quote directly from it and link your accomplishments to its statements to show how you met its requirements. Even better, especially for promotion, quote from +1 or +2 levels *higher* than yours and demonstrate that you're starting to work at those levels.

Not every organization or role will have a career ladder; they are more common in large organizations. If your organization doesn't have one, approach your manager and human resources about starting one!

14.1.3 Tracks: Individual Contributor and Manager

In many tech and engineering organizations, moving up does not necessarily imply becoming a manager. Instead, there are commonly *tracks* for individual contributors (ICs) and managers that—in theory—extend in parallel to the highest possible levels. Thus, an employee could be promoted to the top 1% or so of employees as either an individual contributor or as a manager.

The goal of these parallel tracks is to reduce the pressure for employees to go into management when it is not the right choice for them. A high-performing engineer or researcher should be valued for work as an engineer or a researcher, even if they are uninterested or unable to serve as a manager. Conversely, this recognizes that others are keenly interested in roles as managers and are well-suited for them. When opportunities for promotion and salary are equalized across ICs and managers, the choice can then be made more appropriately for each person.

Table 14-2 shows an example (based on Google when Chris and Kerry worked there) of job titles for ICs and managers, aligned by level. The UX Manager role begins at L5; before that, there are only IC roles. (Note that this varies by discipline. For example, management for PMs might begin at a higher level.)

Table 14-2. *Levels for UX Management vs. Individual Contributor*

Level	Manager Title	IC Title
L4	*not applicable*	UX Researcher III
L5	UX Manager	Senior UX Researcher
L6	UX Manager II	Staff UX Researcher
L7	Senior UX Manager	Senior Staff UX Researcher
L8	UX Director	Principal UX Researcher
L9	Senior UX Director	*(rare)* Distinguished Researcher
L10	Vice President, UX	*(extremely rare), unique titles*

As always, the specific job titles and their alignment to levels varies from one company to another. Table 14-2 is primarily an illustration of how IC and management tracks can parallel one another.

The choice to go into management or to remain an IC is not a permanent choice. Many UXers (and PMs and engineers) switch between the tracks over time. A common pattern is to go into management for a while, perhaps accumulate a promotion or two, and then switch back to an IC role.

14.1.4 Distribution of Levels

Levels are not uniformly distributed. As a Quant UXR, you probably expect that (although we still wish to emphasize it). The linear nature of levels' names (L3, L4, L5, etc.) misleads people to think of them as a simple progression that employees ascend in some orderly fashion. In reality, the frequencies more often resemble a Gaussian distribution, where the highest levels are rarely achieved and most employees spend most of their careers at levels that are labeled as being in the "middle."

This should not be perceived as a negative! The levels that nominally appear to be in the center—such as L4, L5, and L6—are, in fact, high points where employees have substantial expertise, high responsibilities, and much autonomy. The engineers, PMs, researchers, designers, marketers, salespeople, and everyone else at these so-called middle levels create the products and services that every customer uses. These contributions are in no way "lower" than those of employees who are labeled with something later in the level sequence.

Figure 14-1 shows a hypothetical distribution of employees at Levels L2–L10, divided into IC and management tracks. There is a high overlap at L5 and L6. Although those are the most common levels for managers, at each of them there are many more ICs than managers. Each of the two distributions is similar to a truncated normal (Gaussian) distribution; the management level starts at L5, while there are almost no ICs at L8 or higher.

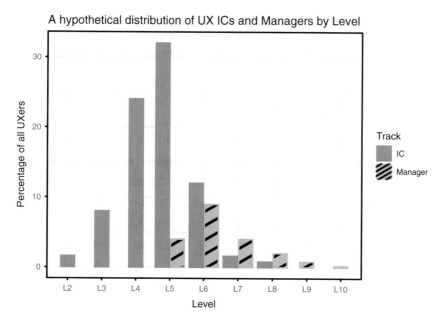

Figure 14-1. *An example of the normal (Gaussian) style distribution one might see for ICs and managers in a UX organization at a large company*

The net result of the situation in Figure 14-1 is to create pressure on ICs to switch to the management track. Although it is theoretically possible to advance beyond L6 as an IC, it becomes more and more difficult; at L7 and higher, the bulk of the distribution switches to management. That leads to a choice in career track, which we discuss next.

14.1.5 The Choice of IC vs. Manager

After a few years as a UX researcher, you can expect to make a choice: should you continue as an IC or try management instead? The choice might seem easy: if there are relatively more managers at the higher levels, wouldn't that be the best track for advancement?

Well, it might be...but only if advancement or being a manager is your primary goal. The counterargument is that management can be more stressful, and—surprisingly to many—may have *less* autonomy and control than being an IC. Put briefly, to take on a management role is to subject yourself much more to the dictates of an organization. A manager is expected to follow, uphold, and promote the official line and impose it on others. An IC can be more independent by, for example, speaking relatively openly about trends, or refusing a given project (although still risking a hit on their performance review).

In our experience the best managers are those who truly enjoy working with people and helping them excel, while encouraging autonomy and ensuring that the ICs get credit, attention, and reward for their work. These managers often end up beloved by those who report to them, and that leads to benefits for everyone.

The worst managers are those who believe that they are smarter than their team, that their job is to tell others what to do, that they deserve the most credit, and that employees must be scrutinized closely to ensure they do not slack. Their teams will resent them and the best employees will transfer out. If this is unchecked, it will lead to an unhappy and poorly performing team. Another common pattern of bad management is one of attention and timing: a busy manager may find it hard to engage fully enough to give meaningful feedback on work in progress, and then suddenly swoop in to micro-manage only when they perceive that something is "wrong."

If you are considering management, think very carefully about your own style and how it will manifest in one of the directions we described. Are you more interested in technical excellence and influence? If so, strongly consider remaining an IC.

We have two suggestions for those inclined to try management. First, ask about managing contractors as an alternative to managing full-time employees (FTEs). There is far more overhead associated with managing FTEs (performance reviews, management trainings, meetings with other managers, etc.). In managing contract ("temp") employees, you may discover whether you like or dislike the people aspects.

Second, collect some data! Assuming that colleagues' calendars are visible to you, review them carefully and compare them to what you prefer doing. Also ask your manager or another manager you trust to go over their calendar with you. Then answer yourself honestly about your response to their schedules. Are their days overwhelming to you? Or are they well-structured or exciting? Do you enjoy it when you have a full day available to code and analyze data? Or would you rather spend your time influencing decisions about budgets, hiring, or performance management?

It is not especially uncommon for ICs to have managers who are at a *lower* level than they are. In Figure 14-1, you can see that there are more L6 ICs than managers; and although each manager will have multiple reports, it would not be unusual at all for some L6 ICs to report to L5 managers (or L7 to L6). It is even more common to have a manager at the same level (both L5 or L6, etc.). This highlights the fact that management should not be conceived as being *over* other employees but rather as being *beside* them and complementing their skills. Both ICs and managers should each boost and support the other.

14.2 The Problems with Levels

The apparent progressive nature of the levels—from L3 to L4 to L5, and so forth—is vastly overemphasized and is too often destructive of careers and happiness. It is quite easy for employees, managers, and organizations to become obsessed with moving "up" the levels.

That obsession leads to three problems that we observe repeatedly. First, it is extremely common for employees—especially when newly hired—to believe that "higher is always better." New employees are often excessively concerned to start at as high a level as possible. This may result in them being thrown into situations where they struggle, and it may create dissatisfaction when promotion to the next level takes a very long time.

A second problem is that it encourages employees and their managers to focus on "moving up" the career ladder. It is easy for a manager to focus on tasks needed for promotion, but that may not be the most important, urgent, or even desired outcome for a given employee.

The third problem is that it becomes more and more difficult to gain promotion as an employee moves up. The influence and impact expected increase (described in the next section), and the effective "distance" between the levels grows larger. A more realistic scheme might label IC levels L4, L5, L6, L7, and L8 as being more like L15, L40, L70, L96, and L99—similar to the cumulative percentiles of Figure 14-1.

14.3 Performance Reviews and Promotion

Performance reviews (which we'll abbreviate as "perf") cause endless anxiety for nearly all employees in tech. These are quarterly, semi-annual, or annual evaluations of an employee's work and usually include written feedback and a formal rating.

Promotion (sometimes just called "promo") is an increase in level, with an accompanying increase in pay and responsibilities. Promotion requires a pattern of strong performance reviews, but is typically a separate process and not guaranteed by the reviews alone.

In this section, we'll look at perf and promo in turn.

14.3.1 Performance Reviews

Performance review ratings are often given on an ordinal scale, perhaps with four to six levels (often five). We show an example of a perf rating scale in Table 14-3 and the proportion of employees who might receive each rating. This scale is similar to ones that have been used at various companies with either text or numeric labels.

Table 14-3. *Example Performance Rating Scale*

Rating	Given to	Notes
Needs improvement	5%	Employee must improve; otherwise, will be fired.
Meets expectations	40%	Work aligns with job description. OK but not outstanding.
Exceeds expectations	30%	Results are better than formal expectations; employee is performing very well.
Strongly exceeds expectations	20%	Outstanding work, may be indicative of next ladder level.
Superb	5%	Uniquely excellent work, contributions far above expectation.

We emphasize two important inferences from Table 14-3. First, very few people receive the highest ratings. It is dependent on the expectations at the current level, and is not the same as getting an "A" in an educational setting. If only 5% of UXers are given a "superb" rating, then you may go for many years—or an entire career—without receiving that label. Don't worry about it! Second, almost everyone receives scores of "meets expectations" or better. That is typical in organizations that hire well and attract great employees; it is perfectly fine to do your job well in the company of many talented colleagues.

How are review scores chosen? It varies by company, although a common model is that the employee self-documents and describes the impact of their work (Section 14.3.2); then their manager proposes a review score to be reviewed by their own peers or manager; and then a final score is awarded. Many organizations have quotas or partial quotas. For example, the organization might require 5% or 10% of employees to receive the lowest possible score, and may limit the sizes of high score buckets.

Many organizations also have a *stack rank* system, where everyone on a team—such as all UXRs or all UXers on a particular product—are ordered from highest to lowest by relative contribution. The stack rank order is then mapped to the ratings according to a forced distribution such as the percentages shown in Table 14-3.

There is substantial research indicating that inflexible systems such as stack ranking and forced distributions are detrimental to employee satisfaction and retention [139]. We encourage teams to emphasize reward rather than punishment. We note the perf systems here not because we recommend them but only to describe what you might expect in industry.

14.3.2 Impact

Performance ratings are usually awarded on the basis of *impact* to the product or business. But what is impact? An honest but circular answer is that "impact" is defined as whatever your colleagues and management *say* is impact.

The most common reflections of impact are changes to a product, strategy, or process, such as the choice of one design over another; the selection of key features and capabilities; influencing a strategic decision to launch a product or not, or to focus on some customer segment; or creating a process, tool, code library, or so forth, that is used by many colleagues or customers. In short, *impact* means a positive change that is attributable to one's actions.

For Quant UXRs, it can be difficult to attribute change to their work because it influences many stakeholders in a diffuse way. Whenever there are objective markers—specific decisions you informed, or measurable outcomes like improvement in user retention—those are helpful to document and include in your performance reviews. You may also have organizational impact, such as defining the key metrics used to measure product success.

However, the most important thing is to make sure that you are in communication with colleagues or management about what they will say about your work. If a PM is willing to write that your work had impact, then—almost by definition, as just noted—it had impact. It is best to ask soon after the relevant work occurs; don't wait until review time many months later. Do not annoy your colleagues by pestering them to say nice things about your work, and yet at the same time, don't be shy about it. The balance between *enough* and *too much* feedback can be difficult to ascertain, and that is a good conversation to have with your manager or a colleague mentor.

14.3.3 Promotion

There are two common models for promotion: the *nomination* model and the *reward* model. In the nomination model, which seems to be most common, you nominate yourself for promotion and document the rationale for it. Then your manager provides their assessment and some separate group (perhaps their manager, or perhaps a promotion committee) makes the final determination.

In the reward model, promotion is simply given to you without a formal nomination or even your awareness that it is being considered. Management observes everyone's work and then awards promotion when they believe it is warranted.

Our view is that the reward model overall works better than the nomination model, because it is less complex and comes with less emotional burden and risk of disappointment on the part of employees. Writing a justification packet requires a lot of time and effort, and it is very difficult not to become emotionally invested in the outcome.

There are a few details about the promotion process that newcomers to the tech industry may not be aware of:

- Promotion does not occur when you are *able* to work at the next level, but when you are *already working* at the next level. The firm will not want to gamble that you could do the work; it wants to know that you are already doing it.

- Promotion requires strong performance review scores, but those are insufficient on their own. You might be performing quite well at the current level but not yet showing work at the next level.

- In addition to review scores, promo needs support from both your manager and stakeholders at higher levels. For instance, there may be an informal expectation that promotion to level X requires support from several folks at level X+2 or X+3. Thus, promo to L5 might require support from two or three reviewers at L7 or L8. Don't be shy about asking for high-level support. (It is, again, a nice feature of the "reward" model that it does not require ICs to do that on their own; managers do it silently for them.)

One very important thing to note, in our view, is that *you don't have to go for promotion* (or accept it, in the reward model). If you're happy where you are, it may be wise to stay there. There are other ways to grow in enjoyment, influence, and satisfaction. At higher levels, the expectations will increase, and the projects and day-to-day work will change—not always in ways that you prefer. Tech roles also typically have reasonable budgets for salary increases within a level, without any need for promotion.

Put differently, a senior or staff UX role or its equivalent is a superb job, and you may find it very satisfying to enjoy that role and not worry about an arbitrary level system, especially if promotion is highly political or would move you away from what you enjoy.

In the nomination model, if you apply for promotion and don't get it—even if you began the process with an attitude of "I'll try it and see"—you are likely to feel very discouraged, probably more so than you would expect. You may find yourself wanting to change projects, teams, or companies as a result.

We don't wish to be discouraging; rather, we hope we are realistic. The benefits of promotion in terms of higher salary, higher respect, and greater influence are all valuable and rewarding. Yet those benefits come at personal cost and higher expectations from an organization. Each one of us must balance those aspects across a career.

14.4 Personal Styles and Goals

Although it is an obvious truism that you should make career choices according to your personal goals, it can be difficult to do that in practice. Our own goals are often opaque to ourselves. They change over time with personal growth and experience, and they are influenced by social and institutional pressures. In this section, we share a few observations that may help achieve some degree of clarity.

14.4.1 Maximizing vs. Satisficing

The Nobel prize-winning political scientist, economist, and computer scientist Herbert Simon described two decision-making approaches that may also be seen as individual psychological traits: *satisficing* and *maximizing* [134]. When people *maximize* a decision—such as choosing a job—they attempt to review all available information in depth, consider every option, and arrive after much study and deliberation at the most likely optimal choice. By contrast, someone may *satisfice* a decision by choosing an option that is "good enough," with awareness that there may be alternatives that are better in some way.

Although no one makes every choice using a single approach, and it is somewhat reductive to speak of types as opposed to traits, for simplicity we might consider the approach that a person tends to use more often in some domain and describe them as a "maximizer" or "satisficer."

In our experience, maximizers are the most common variety of Quant UXR. So, if that sounds like you, you are likely to be in good company as a Quant UXR (and, in fact, in tech research generally). In this case, career satisfaction is likely to be achieved by having a variety of new challenges that afford the opportunity for deep learning and skill development. This may be particularly reinforced by situations where you can demonstrate mastery, such as building shared tools, teaching, or publishing. On the other hand, it will be detrimental to deny those possibilities, as might occur in a job that relies on situations that are rarely mastered (such as organizational politics) or that continually change and afford no potential for depth (such as superficial analysis and reporting). When you consider a job, think carefully about the day-to-day tasks and whether they will fit your personal style.

There are two common risks for maximizers. First, management may not understand their motivation. This may lead to managers directly opposing activities they thrive on (for instance, by asking them to stop learning new methods, teaching, or publishing) or by pushing them to engage in things they will not enjoy (such as opinion-driven meetings or higher velocity but repetitive work). Second, they may simply run out of projects that interest them in a particular role.

For satisficers, UX research positions are also outstanding career choices. The positions tend to be as stable as any industry positions, are in demand, and are well paid. They also bring personal reward from interacting with exceptional colleagues, working on interesting products, and improving technology for users. The work itself is predictable when a researcher uses standard methods. UXRs in senior roles (L5) tend to have an enjoyable level of autonomy—they are sufficiently senior that they are trusted to run a research program, yet not so senior that organizational demands become a distraction or burden relative to IC work as a researcher.

14.4.2 Builder vs. Explorer

There is another personal style that largely cuts across the dimensions of satisficing and maximizing. We refer to it as *building* vs. *exploring*, following the philosopher T.K. Seung [130], who applied the notion to philosophers and other thinkers (Seung used the term "settler," but we prefer "builder.")

A *builder* stays with a particular set of problems while evolving methods and approaches over a span of decades. To stay with Seung's field of philosophy momentarily, we could say that builders include philosophers such as Aristotle, Thomas Aquinas, Immanuel Kant, and more recently John Rawls. Although thinkers such as Aristotle may cover vast territory, the similarity in their work over time is unmistakable. Builders may create grand, systematic structures.

Explorers move among problems and methods according to changing interests. They tackle one set of problems, with one approach, and then move to something quite different, and may be unconcerned with continuity of topic or method, or even consistency over time. The philosophers Plato, Friedrich Nietzsche, Ludwig Wittgenstein, and, to a large extent, Seung himself fall into this camp. Explorers do not attempt systematic syntheses, or when they do, as with early Wittgenstein, they may abandon them for a quite different style later. Explorers tackle one set of problems deeply and then move to something quite different.

In the Quant UX space, a builder might become increasingly expert in an area such as customer satisfaction research or logs analysis. We have discussed T-shape skills a few times (Section 4.1.1) and would note that it is not contradictory to "settling" into a specific area. There is tremendous breadth as well as depth available in any area of Quant UX, whether that is survey research, A/B testing, logs and behavior analytics, customer satisfaction, choice modeling, or any other area.

A Quant UXR explorer moves across quite different projects over time. At one time, an explorer may do deep logs analysis while at another time they may tackle a psychometrics project or do qualitative research. Their breadth expertise is in research design *in general*, supplemented by depth in a few and perhaps disparate methods.

You might satisfice or maximize as either an explorer or a builder. A satisficing explorer would be happy with "good enough" work in any particular area before moving to the next. On the other hand, maximizing explorers passionately school themselves to "learn everything" in a new area; then they may abandon it for the next topic that comes along. A maximizing builder would develop great depth of knowledge in an area, perhaps becoming a world-class expert in it. A satisficing builder would be content to do their job well, delivering useful results and enjoying the work, without concern for comprehensive understanding of every detail.

Any of these styles is perfectly acceptable as a Quant UXR—or as a general UXR, designer, engineer, PM, or any other tech role. Our point is that you should understand your own style so you can set expectations and make choices that align with it, paying more attention to your own needs and skills rather than messages and pressure from others.

14.5 Building Skills Throughout a Career

You'll want to build skills over the course of a career and yet there is a problem: at the beginning, it is a *multi-armed bandit* problem [135]. There are many levers you could pull—building skills in statistics, programming, qualitative research, project management, people management, and so forth—but you don't know the expected payoff for any of them.

Our suggestion is the same as for solving the technical multi-armed bandit problem: try several paths and observe the outcomes in terms of your own assessment of what matters to you. Devote progressively more effort to the areas that are rewarding, but do not place all of your efforts there; continue to devote some effort into ongoing learning in other areas. That strategy ties well into building out T-shaped skills (see Section 4.1.1). As your skills grow, take care to continue adding both breadth and depth.

14.5.1 Areas for Skills Development

With the multi-armed and T-shaped strategies in mind, we have some observations and suggestions in several areas.

14.5.1.1 Quant or UX?

Relatively early in a Quant UX career you will likely incline toward greater interest in either the *UX* side of Quant UX—designing and making products better for users' needs—or the *quant* side, being more interested in the technical aspects of statistical modeling or programming. We strongly recommend that you not decide this too early, and develop these skills in parallel.

On the other hand, if you incline strongly and inexorably to the quant side, then you may eventually wish to transition to a data science or similar role (see Section 3.4). Movement between roles in tech is generally easy, if you have the necessary skills.

14.5.1.2 Qualitative Research Skills

Too many Quant UXRs are afraid of qualitative research. Developing comfort with qualitative engagement such as user interviews will help both your quant projects and your understanding and relationships with other UXers. This will benefit both your immediate work as a Quant UXR IC and establish understanding of general UX research that will be valuable if you later explore general roles or become a UX manager with general UXR reports. The first area of qualitative breadth to explore is this: learn to do *individual depth interviews* (IDIs) with an emphasis on *Think-aloud* protocols [42].

Think-aloud protocols are applicable not only to traditional usability studies [44] but also when pre-testing quantitative studies, especially survey research. If you enjoy IDIs, you may wish to continue to learn about usability lab studies [99], general interviewing [108], field research such as home and site visits [78], and focus group research [73].

14.5.1.3 General UX Research…and UX Generally

Throughout this book we have emphasized that Quant UXR roles are a part of UX research positions in general. The more you know about UX research and UX organizations, the better you will engage as a Quant UXR.

We have reading recommendations in Section 14.8, but the two most important recommendations to build UX skills are these: run a variety of studies using qualitative methods as well as quant methods, and spend time learning what your UX counterparts do.

An excellent way to start is to ask a general UXR colleague if you can shadow some of their research. Accompany them to meetings with designers and PMs to plan the research. Volunteer to take notes for them as an observer during a usability lab study or field research project. Contribute to their findings report (if they wish) and attend any presentations. And then turn the table: ask them to assist you while *you* lead the next round of a general, qualitative study. Even if you decide not to do qualitative studies often, the learning will be immense—and you'll have a colleague to write something nice on your performance review, noting a project with unusual breadth for a Quant UXR.

14.5.1.4 Programming

The question about skills development in programming is relatively straightforward. Are you able to program at an intermediate level? If not, we strongly recommend to develop the skills described in Chapter 6, "Programming."

If you already do program at an intermediate level, then we recommend to consider building statistics skills instead of stronger skills at programming. Incremental knowledge of statistics has a larger and more general payoff for Quant UXRs, and it is more unique within tech companies (who have many experts in programming).

If you wish to build advanced skills in programming, we suggest giving specific attention to collaborative coding (Git or whatever your organization uses); notebooks and other team-based data analysis platforms (Jupyter, RStudio/Posit notebooks, Observable, etc.); unit testing; package and library development; distributed computing (i.e., cloud computing); and computational statistics.

14.5.1.5 Statistics

As we just mentioned, learning more about statistics is the single best investment of time in skills development for many Quant UXRs. If you know little about statistics, learning more will rapidly enable you to undertake a vastly larger array of projects; or if you know much about statistics, you will discover new tools, methods, and approaches that magnify your impact and enjoyment.

We give specific recommendations for advanced learning about statistics in Section 5.6.

14.5.2 Find Mentors

Many companies have formal mentorship programs, and you may be assigned a mentor as a new employee. Whether that relationship works well is largely random; some mentors and mentees do not click, while others have long-lasting, mutually beneficial engagement.

Regardless of what your organization may prescribe, we recommend that you seek out mentors to advise you in three areas:

- Day-to-day organizational issues such as how to schedule meetings, book conference rooms, recruit users, find research suppliers, write supplier contracts, plan travel, and file expenses. This mentor may be a general UXR colleague or, if you have one on your team, a UX program manager. This relationship may be sporadic and ad hoc and yet hugely helpful.

- Technical research issues such as helping you work with the organization's code repository or interpret statistical models. Never hesitate to ask a colleague to show you how to do something. Every large organization has complex systems that can only be learned through undocumented, collective knowledge that is shared person to person.

- Career issues such as your day-to-day happiness on the job, how to write performance reviews, and how promotion works.

All three of these areas might be covered by a single mentor, but more likely they would span two, three, or even more relationships. The best way to find someone is to observe your colleagues' interactions and then simply to ask them. There is no need to go through a formal mentorship program (although that is also OK).

Every mentoring relationship should start with a discussion of boundaries and expectations for privacy. You do not want a mentor who is secretly reporting on you. For our part, we always call out specifically that mentoring is confidential (up to the full provisions of the law and employment requirements). In our view, mentoring is all about helping the other person. In return, the mentor learns from the person being mentored and should not attempt to accommodate some organizational agenda.

14.6 Paths for Senior ICs

Until very recently, there have been few Quant UXRs at staff or higher levels, simply because there have been few Quant UXRs. Even among software engineers, the career trajectories above senior (L5 in our scheme in Table 14-1) are difficult to characterize in general. In his book *Staff Engineer* [80], Will Larson comments (p. 10) that "There is pervasive ambiguity around the technical leadership career path [for engineers], making it difficult to answer seemingly simple questions about Staff-plus roles."

When the situation is unclear for engineers—who may outnumber *all* UXRs at a large company by 10:1, 20:1, or more—it is even murkier for Quant UXRs. Thus, one point we want to emphasize is this: you're not alone if the paths are confusing to you. There is a positive corollary: you can help establish the paths. (Managers, take note! Staff and higher positions for Quant UXRs present an opportunity for creativity, evolution, and growth of the role. Don't expect your senior ICs to follow the prescriptions of a career ladder or the paths taken by designers or engineers.)

It is also common for Quant UXRs to move out of the role altogether. Among the Quant UXRs that we have known for 10 years or longer, few are still in Quant UXR roles. Common choices for their next roles have included general UX researcher, UX manager, data scientist, developer, freelance UX researcher (consultant), and PM. Although it is wonderful to see those options, we also hope that as the field develops there will be more opportunities to grow and stay within Quant UX.

The movement of Quant UXRs into other roles is not necessarily due to limitations on upward growth in Quant UXR; it is also due in large part to the nature of Quant UX as an intersection of multiple skill sets (see Section 4.1) and how that relates to depth and breadth knowledge (Section 4.1.1). Someone who has all of the intersecting skills—statistics, programming, and UX—will naturally develop more depth over time in one area, and that may lead them to move to a role where that is emphasized. For example, after learning JavaScript to create the original sequences sunburst visualization (see Chapter 9, "Log Sequence Visualization"), Kerry wanted to build deeper skills in JavaScript programming, and went on to work as a data visualization developer for several years.

Despite the lack of clarity in very senior roles, there are three general paths that we have observed emerging as patterns for Quant UXR ICs at the level of staff and above: the *tech lead*, the *evangelist*, and the *strategic partner*. These do not reflect job titles but describe possible areas of emphasis for very senior researchers. The patterns are not strictly separate and may overlap in practice; this could be good or bad depending on whether it represents synergy as opposed to conflicting goals. Still, we believe that it is helpful to emphasize one of them and set expectations carefully. Attempting to cover all of the patterns simultaneously and deeply is unlikely to succeed.

In the following sections, we comment on each pattern and outline its advantages and drawbacks.

14.6.1 Staff Level Pattern 1: Tech Lead

A Quant UXR *tech lead* (TL) takes on complex projects that are large, long range, and of particular importance. Sometimes these are individual projects, but more commonly they involve coordinating, synthesizing, or laying technical foundations for the work of other Quant UXRs. A Quant TL may develop a research approach and tools to solve a specific problem at hand and then make them available to other Quants or general UXRs for reuse when facing similar challenges in their own projects.

There are three examples of this kind of work in this book. First, Kerry led the early Quant UXR team at Google, and recognized the need for a focus on UX metrics; this resulted in the HEART framework, which became widely used at Google and elsewhere (see Chapter 7, "Metrics of User Experience"). Second, Chris has developed internal and external tools for choice modeling and the assessment of customer needs. His MaxDiff tools have been used by hundreds of UXR colleagues (see Chapter 10, "MaxDiff:

Prioritizing Features and User Needs"). Finally, Kerry's sequences sunburst visualization offers an elegant solution to the common problem of representing aggregated user paths through a product, and has been reused by teams across the tech industry (see Chapter 9, "Log Sequence Visualization").

The advantage of the tech lead pattern is that it lies very close to the core technical skills for Quant UX research. It is what many Quant UXRs imagine and hope to do as they advance in their careers, and it is also the most common pattern for a UXR at the staff-plus level.

However, that closeness to core Quant UXR work is also its largest disadvantage. It may be difficult for non-expert stakeholders to understand the need for advanced methods and approaches, or why a particular problem is difficult rather than easy. Quant TLs may go unrewarded for solving difficult problems while also being asked to do mundane work that does not interest them.

Another risk for a tech lead is being asked to take a dual role that combines management alongside IC work. This is often pitched as a win-win situation that brings all the advantages of clearer upward mobility from the management track with the opportunity to keep doing IC work that the TL loves. Our view is that this often does not turn out as expected. It frequently becomes a situation in which either the IC work is abandoned under the pressures of immediate management or the TL is required to work much longer every week. Before accepting a dual role, talk with several others who have similar positions, and—as we noted previously—inspect their calendars.

The most important requirement for success as a TL is to have management and stakeholders who understand the difficulty of the work and the value of the problems you are solving.

14.6.2 Staff Level Pattern 2: Evangelist

The *evangelist* pattern is closely related to that of a tech lead, and often one person takes on aspects of both. Whereas a TL builds systems, leads large-scale efforts, and solves problems, an evangelist takes those solutions to others. An evangelist is responsible for bringing best practices in Quant UX research to the larger organization, whether that is other researchers or stakeholders.

The primary difference from a tech lead is in the relative attribution of effort and the way that impact is judged. A TL will be evaluated by the difficulty of the problems and the value of their solutions; an evangelist will be evaluated for influence on others.

The particular problems tackled by an evangelist may not be especially difficult in themselves; the value of the role comes in sharing them across an organization.

An example related to this book is the Customer Satisfaction (CSat) program that Chris initiated in a Google division that had never had one (see Chapter 8, "Customer Satisfaction Surveys"). The value of the effort was not in the specific difficulty of any single project but in the experience and effort to build out, share, promote, and demonstrate the practices so they were widely adopted.

Advantages of an evangelist role are that it draws on a large range of skills, including technical work and educating others. It is directly rewarding to interact with a large group of colleagues and see the impact you are having on their work. The primary disadvantage of an evangelist role is that its contributions are diffuse and are easily devalued by the management of the evangelist's own team, especially when there is turnover or reorganization at a management level. It may also become dull or repetitive for a Quant UXR who is inclined to explorer or maximizer styles (see Section 14.4).

The primary contributor to success in the evangelist pattern is unfortunately something that is largely outside of one's control or choice: the cooperation and expertise of others. You may be unable to influence others for many reasons. Their part of the organization may have differing goals; they may view your work as desirable but less important than other projects; they may have aspirations or plans that directly conflict; there may be disputes at an executive level that produce conflict; and so forth. Some companies have cultures that are much more amenable to cross-company collaboration and influence than others. Assess the situation in advance so you will have a good idea of the possibilities, and to the extent possible, scope out some likely "easy wins" to get started.

14.6.3 Staff Level Pattern 3: Strategic Partner

The *strategic partner* pattern arises when a very senior UXR takes on a project of very close attention and high value for an executive stakeholder (most commonly a VP, but sometimes higher-level executives). The strategic partner leads research to address a large-scale and long-range question of importance, such as "How much risk does our business face from [some situation]?" or "How should we change in the next 5 years to adapt to [some trend]?" or "What should be our next big product investment?"

The advantage of strategic partner research is that it is exciting, ego-gratifying, and promises enormous forward-looking impact. You have the direct attention of an executive, and that conveys both immediate credibility and strong support down the road for performance reviews or promotion. You may be able to secure a substantial budget or headcount needed to implement your research plans. It may promise long-term advancement such as being the director of research in a new product division. And, to top it off, you quite likely will be free of the day-to-day concerns of making and shipping a product.

The concerns are almost a mirror image of those advantages. We'll start with the separation from product engineering; in any tech company, that is a huge risk because the business ultimately depends on successful products. If there is a business downturn, or even a simple change in management, such special projects are often cut first. Close executive attention comes at the cost of high demands, stress, and potential misunderstanding. Executives often expect results that are both faster and more conclusive than research can deliver. There is also the question of temperament. You may only ever have seen the best qualities of the executive, but in working with them closely, you will learn how they behave under pressure. Are they a jerk or a narcissist? If you deliver bad news about a pet idea, what will be the response? (See Section 4.5.5.)

This is a time to be honest with yourself about your own style and anxiety. Does it excite you to take on such a large and critical task? Do you feel confident about your skills to tackle it? Would you really want the company to make a huge bet on your own recommendation? Or will such pressures worry you endlessly as you fear making a mistake or being shown up by someone else?

Every strategic partner situation differs, but we will note one thing: if you don't personally *like* the executive to whom you would report, and if you don't also believe that the executive has a solid, albeit non-technical, grasp of *research*, then such a role may be disastrous.

14.7 Key Points

The Quant UXR role is new enough that every one of us has enormous potential to define, shape, evolve, and change it. When Chris was at Google, he worked in five different product teams in 10 years—and in four of those five teams, he was the first Quant UXR they had ever had. Kerry helped to originate the Quant UXR role in the first place, and that initial team took on early projects that became a basis for the work of many others.

The Quant UXR journey in tech companies and other industries and organizations is just starting. In this chapter, we have reviewed these points:

- Large companies have career levels that express the trajectory of advancement, although in very generic terms (Section 14.1.1).

- There may be a career ladder that defines job roles and responsibilities, relative to the career levels. Career ladders are usually vague as to the *quant* aspects of Quant UXR (Section 14.1.1), but you can refer to them for performance review, promotion, and general career discussion.

- Many organizations have parallel tracks for individual contributors (IC) and managers. You may advance to a very high level as an IC without needing to become a manager (Section 14.1.3).

- The first choice between management track and IC is made around the mid-career point ("senior researcher"). It is not a permanent choice, and many UXRs switch between the tracks over time (Section 14.1.3).

- The uniformly linear labels used in most job-level schemes are misleading. Levels are not uniformly distributed, and there are few UXRs at the most senior levels (Section 14.1.4). It can be a superb and rewarding career choice not to pursue additional promotions at some point, and instead to remain around the "senior" level (Section 14.3.3).

- The most important thing for performance reviews and promotion is your impact. "Impact" is defined circularly as whatever your managers and peers *say* is impact. The crucial point for Quant UXRs is that you will be evaluated mostly by the results you deliver and how useful those are to others, not for the technical sophistication of your methods (Section 14.3.2).

- In sorting out your career goals, we suggest you consider your own tendencies with regard to maximizing or satisficing, and exploring vs. building (Section 14.4). Any of those styles is compatible with Quant UXR but interacts with choices about careers.

- When developing skills over the course of a career, pay attention to the T shape for skills; work on both breadth and depth. You may do so in several areas according to your interests (Section 14.5.1).

- For very senior UXRs—at the level of staff or principal—there are no road maps and every role is largely unique. We identify three patterns to consider: tech lead, evangelist, and strategic partner (Section 14.6).

14.8 Learning More

Our recommendations to learn specific skills are collected in the "Areas of Skill Development" section of this chapter (Section 14.5.1) and in the "Learning More" sections of other chapters in this book, especially Chapter 5 for statistics (Section 5.6) and Chapter 6 for programming (Section 6.6). The "Exercises" section in the technical chapters will cement your learning.

For general career learning, we have a few recommendations. We have emphasized the UX aspects of Quant UX, and learning more about general UX methods is helpful for those new to industry. For general, so-called qualitative UX methods, we recommend Baxter, Courage, and Caine, *Understanding Your Users* [4]. A more general book on how UX organizations approach user-centered design is Krug, *Don't Make Me Think, Revisited* [74].

The day-to-day practice of UX research is the topic of Bernstein's *Research Practice* [6]. It collects stories and reflections from dozens of UX researchers about how to enter, succeed, and grow as a UX researcher. Although it is not specific to Quant UXR (that's why we wrote *this* book), most of the topics are relevant to Quant UXRs.

With regard to personality styles, if you're reading this, there's an excellent chance you are a maximizer or an explorer, or both (Section 14.4). In that case, we recommend Sher's *Refuse to Choose!* [131], which is devoted to discussing careers for those Sher calls "scanners"—who are very similar to what we call explorers. It will provoke you to consider several styles of how to relate to careers.

For more senior UXRs, we have three recommendations. If you're interested in UX management, see Lund's *User Experience Management* [86]. And to learn more about IC opportunities above the senior level, it is highly worthwhile to read Larson's *Staff Engineer* [80]. Although Larson is concerned primarily with software engineering, the text is non-technical and most of its reflections and recommendations apply to staff-plus positions in UX research as well as engineering. Reilly's *The Staff Engineer's Path* [112] is especially helpful for its discussion of the individual challenges, changes, and rewards that one faces at higher levels. It is highly applicable to UXRs as well as engineers.

For general career advice, we also recommend Robinson and Nolis, *Build a Career in Data Science* [114]. Although it focuses on data science rather than Quant UX, the chapters on the community, working with stakeholders, applying for positions, and—to some extent—technical skills are all relevant to Quant UXRs.

You will find hundreds of Quant UXRs meeting to share research presentations, engage in skills workshops, and discuss careers at the annual Quant UX Conference. Quant UX Con is an emerging community where your contributions will be welcomed and formative of the future. Details are at `https://quantuxcon.org`.

## CHAPTER 15

Future Directions for Quant UX

Where will Quant UX go in the future? As we have observed throughout this book, Quant UX does not have a precise or static definition; it is evolving as all of us create and expand the role. In this concluding chapter, we describe four possible futures we imagine for Quant UX, based on current trends as of 2023.

These four directions are not mutually exclusive, and it is almost certain that each of them will be present in the future. Yes, as the field grows and changes, the relative mix among them is certain to change. (As the science fiction writer William Gibson put it, "The future is already here. It's just not very evenly distributed." [50])

15.1 Future 1: UX Data Science

The first, strong possibility is that Quant UX will be increasingly staffed by data scientists working inside UX organizations. This change would deemphasize human subjects research and elevate technical skills such as machine learning and programming.

This path is perhaps the most likely of the four futures because it responds to several forces: the need for candidates to fill Quant UX positions, the much larger size of data science as a discipline, the number of data scientists available to fill roles, and the general if uninformed impression among stakeholders that data science skills are the most crucial technical abilities for Quant UXRs.

The problem with this path is that it might lose the distinctive human-centered aspects of Quant UX. Data scientists usually have little or no training in human subjects research (see Section 3.4.5), and they often emphasize machine learning whereas we believe a more natural alignment for Quant UX is with statistics (see Section 5.1.1). A counterargument is that data science is already so large and nebulous that it includes all of these possibilities without conflict.

C. Chapman and K. Rodden, *Quantitative User Experience Research*,
https://doi.org/10.1007/978-1-4842-9268-6_15

Our recommendation here is primarily to hiring managers: don't assume that data scientists are interchangeable or that they will adapt to the goals of a UX organization. Assess candidates carefully and look for evidence of human subjects or UX experience.

For individual Quant UXRs, you may wish either to develop stronger technical skills that would align with such a shift (if it occurs) or to consider moving to a different role such as mixed methods UX (see Section 15.3). For data scientists transitioning into UX, we recommend that you develop additional knowledge of human subjects and human-computer interaction research, as needed.

15.2 Future 2: Computational Social Science

Another possible future would see the human-centered aspects become the core of Quant UX. Quant UX would emphasize human subjects research, cognitive psychology, behavioral research, and survey science, while technical areas that have greater overlap with data science (such as logs analysis and A/B testing) would be covered increasingly by data science teams. Quant UX would increasingly reflect the approaches of computational social science, as practiced in psychology, sociology, political science, and other disciplines.

This future might result from the interplay of two forces: first, a desire to clarify the relationship of Quant UX and data science without either of them subsuming the other; and second, the high demand for Quant UX candidates being partially met by social scientists migrating from academia, government, and consulting.

This direction would represent a contraction of Quant UX in some areas along with growth and acceleration in others. It could lower or eliminate the requirement for skills in programming, and it might increase the emphasis on statistics while shifting its center away from common inferential statistics. There would be greater emphasis on latent variable analysis, graph models, structural models, stratified and matched sampling, and similar methods that are often used in social sciences.

This could be a stable and valuable direction for Quant UX, but it is likely to be limited by the availability of candidates with depth in computational social science. Our recommendation to hiring managers is to hire these kinds of candidates when you can find them because these skills are valuable and relatively uncommon. For social scientists, it won't surprise you that we recommend consideration of Quant UX as a potential career.

15.3 Future 3: Mixed Methods UX

Another possible path is the increasing adoption of advanced statistical procedures and programming by general UX researchers. This could come from the growth of the "mixed methods" UX research role (see Section 3.4.2), such that Quant UX becomes dispersed along the continuum of skills of General UXRs.

This future could arise due to the success of Quant UX along with expansion of the role of UX research. UX teams increasingly include researchers with depth in survey research, psychometrics, ergonomics, and many other areas that are each "quantitative" in ways that differ from Quant UX as defined in this book. Rather than attempting to carve out a single specialty, it would be natural to allow each of these specialties to coexist within UXR.

The challenge to this path is that it is difficult to manage and assess quantitative research alongside qualitative research. They involve different training, skills, research processes, stakeholder engagement, and modes of impact on products. It is difficult to hire, manage, review, promote, and generally satisfy members of a team who have widely disparate sets of expertise. It will be a challenge to determine a reward model that recognizes the many different ways in which each discipline has impact, and the result is likely to be tension and a gravitational pull toward a subset of the specialty areas.

In this model, individual UXRs may be encouraged to use methods without adequate preparation or support. For example, a general UXR without training in statistics may use and misuse complex statistical models; or a Quant UXR may run qualitative research without appropriate consideration of the nuances of small sample research or ethical concerns of live research.

On the other hand, with proper attention, management, and a reward structure that encourages collaboration, we believe this model is very promising. It allows a diverse mix of skills without requiring each member to be a generalist, and it enables a large array of projects while preserving the most important aspect of UX research, *focus on the user* (see Section 4.2).

15.4 Future 4: Quant UX Evolution

The fourth future path would be the ongoing evolution of Quant UX as a distinct discipline within UX research. In this future, Quant UX would continue to grow and develop its own identity.

The two trends that push in this direction are the high and increasing demand for quantitative analyses from UX stakeholders, and increasing pressure from Quant UXRs and their managers to clarify the role. Some organizations have historically hired Quant UXRs into the same formal role as all other UXRs, but this has led to confusion when there is no defined career ladder or guidance on the unique differences of Quant UXRs (see Section 14.1.2). Today some of those organizations are exploring Quant UXR–specific career ladders, levels, performance review standards, and all the other elements necessary to make Quant UXR a unique position instead of being a "flavor" of general UXR.

This path could be expected to lead to greater differentiation of roles within Quant UX, allowing differences among logs analysts, social scientists, survey scientists, and others who may share poorly defined roles today (see Section 3.3). This could help quant teams educate their stakeholders, resist the pressure for every Quant UXR to be a generalist, and clarify when specific skills such as programming and advanced statistics are required.

We would also expect it to lend clarity to the promotion process, and eventually to create well-defined Quant UX manager roles. Unlike the dispersion of methods that we imagined in the mixed methods future (Section 15.3), there would be greater concentration and expertise specific to quantitative UX research.

We believe this future would be ideal, as it preserves the distinctive nature of Quant UX while allowing for a greater array and diversity of skills. It would also address some of the most crucial questions that Quant UXRs routinely face about defining their impact, differences from other roles, and long-term career progression.

However, several countervailing forces create difficulties for this path. In particular, it would be challenged by the overwhelmingly larger size of the data science discipline, lack of stakeholder understanding of the differences between Quant UXRs and other roles, the fact that quantitative methods are not clearly separable from other UX methods, and the intense pressure that UX hiring managers face to hire candidates with statistics and programming skills at the expense of experience in UX or human subjects research.

Our recommendation is this: try! As we have argued throughout this book, Quant UXRs bring distinctive contributions that are not easily replicated by other roles. Those advantages are large enough to justify a sustained investment in individuals acquiring Quant UX skills and organizations building Quant UX teams.

The risk in pursuing this path is low. If one of the other futures prevails and Quant UX continues differently than we hope, there will be little downside. Because Quant UXRs overlap with general UXR, data science, and other roles, the contributions of strong individual contributors will always be valuable even if the exact name, nature, or organizational location of the role changes.

15.5 Learning More

The best way that you can learn more about the future of Quant UX is to get involved and help invent it. We hope this book has inspired you and will continue to be useful on your journey.

Apart from joining an organization and becoming a Quant UXR, or using the approaches and methods here in your work in another role, you can find others who are actively engaged in defining the future of Quant UX at the annual Quant UX Conference, `https://quantuxcon.org`.

15.6 Finally

If you've read the book this far, or if you've only read this chapter, or if you jumped to the last page to see how the plot ends…well, thank you for reading! Our most important note about the future of Quant UX is this: *you* will help build it—and we look forward to meeting someday.

APPENDIX A

Example Quant UX Job Description

Following is an example job description for a quantitative UX researcher. We write from the perspective of a UX manager at the fictional company in the Looney Tunes *Road Runner* series, ACME Rocket-Powered Products [83].

If you're a potential Quant UX candidate, read our notes in Chapter 12, "Interviews and Job Postings," on how to interpret job listings and their lists of required skills. Remember that not everything listed in a job description is an actual requirement.

If you're a hiring manager, feel free to reuse any part of this description and update it to remove the rockets (unless you're making rockets!).

Quantitative User Experience Researcher

At ACME Rocket-Powered Products, we know that a great user experience is the foundation of business success. As a Quantitative User Experience Researcher (Quant UXR), you will work with qualitative researchers, UX designers, product managers, and rocket engineers to ensure that we continually focus on improving users' lives through rocketry.

ACME Quant UXRs plan, field, analyze, and report research to understand user needs in depth. You will answer questions such as: What do users want? Which features of rockets do they use? How does feature usage relate to users' characteristics and other aspects of their rocket experience? How happy are users with ACME rockets? What should we do to deliver better experiences and expand the usefulness of rockets? And, for all of these questions, *Why*? As a Quant UXR at ACME, you will engage with qualitative research colleagues to deliver an integrated, quantitative + qualitative understanding of users.

© Chris Chapman and Kerry Rodden 2023
C. Chapman and K. Rodden, *Quantitative User Experience Research*,
https://doi.org/10.1007/978-1-4842-9268-6

If you have a passion for user research, quantitative analysis, and rockets, ACME would love to hear from you. Please apply online by clicking below.

Responsibilities

- Design, field, and analyze quantitative research projects to assess users' needs, preferences, and experiences with rocket-powered products.

- Define and measure user experience metrics for user satisfaction, task success, and other aspects of interaction with rockets.

- Work with designers, qualitative researchers, and product managers to prioritize research needs for new and existing rocket-powered products.

- Present research findings and recommendations to stakeholders including astronauts, designers, product managers, engineers, and executive leaders.

- Identify new research opportunities to improve users' lives through rocketry.

Minimum Qualifications

- A bachelor's degree in a research or related field such as psychology, computer science, human-computer interaction, cognitive science, or others, or equivalent work experience.

- Demonstrated expertise in experimental research design for human behavioral data.

- Ability to perform statistical analysis and write relevant programming code to process human behavioral data.

Preferred Qualifications

- A master's or doctoral degree with a quantitative research emphasis in any related field, such as psychology, computer science, human-computer interaction, cognitive science, and other quantitative social sciences.

- 4 years of experience conducting UX research in a product design team.

- Proficiency in a computational or statistical programming language, such as R or Python.

- Broad knowledge of statistical analysis methods including descriptive statistics, linear regression, multivariable analysis, and statistical inference.

- Expertise in survey design, logs analysis, and A/B testing.

- Strong communication and presentation skills.

- Expertise in communicating complex research to stakeholders from design, product management, engineering, and business leadership.

- Experience in using, building, or researching rockets.

Example Quant UX Hiring Rubrics

In this appendix, we outline a set of possible rubrics to assess the potential fit between a Quant UXR candidate, the example job description in Appendix A, and the skills we describe in Chapters 4, "UX Research," 5, "Statistics," and 6, "Programming." They reflect the authors' opinions and are not derived from any company or set of industry criteria. These rubrics represent what we hope to see in Quant UXR candidates, and we believe our "Minimum" section in each skill category provides a reasonable baseline standard.

For individual applicants, these rubrics should give you a sense of how hiring committees might evaluate your skills, although the exact requirements for any organization or position will vary. Do not expect an organization to share anything like this list of requirements; if one exists, it would be used internally only to assist with discussion rather than as a strict evaluation.

In reading this set of rubrics, the skills listed in each "Minimum" section are ones that we would regard as strictly required for every candidate. Each skill listed in the "Moderate to Advanced" sections is optional for candidates in general, although some of them would be required for any specific opening. A candidate would be expected to show "moderate" or "advanced" proficiency in at least *one* category in alignment with the T-shaped distribution of skills for Quant UXRs (see Section 4.1.1).

Don't worry if you meet the minimum skills but only few of the moderate or advanced ones. Some of those skills may take decades to develop and are not generally expected of candidates who are new to Quant UX.

These criteria should be adapted appropriately to the job level. For example, for an entry-level position such as L3 (see Section 14.1.1), the requirements for research planning would have a narrow scope and short timeline, whereas a senior candidate (L5) would be expected to consider a broader, longer-term, more-comprehensive approach.

© Chris Chapman and Kerry Rodden 2023
C. Chapman and K. Rodden, *Quantitative User Experience Research*,
https://doi.org/10.1007/978-1-4842-9268-6

For hiring managers, we hope this list will help you create a set of rubrics that fit your needs and organization. Feel free to adapt or reuse any parts of it.

Rubrics to Assess Quant UXR Candidates

Skill: UX Research (Chapter 4)

Minimum

- Designs research from a user-centered point of view, considering data sources and measurements that relate to user experience.

- Understands principles of sampling, sample bias, and reproducible research.

- Able to create appropriate research plans for A/B tests of UX designs, and similar studies.

- Coursework, training, or job experience with human subjects research.

Moderate to Advanced

- Defines research plans with multiple components, such as planning iterative A/B tests paired with logs analysis or survey research.

- Identifies sequences of research projects to occur longitudinally, to address product needs more completely and holistically.

- Clarifies the larger picture of how a research request aligns with broader user experience goals.

- Demonstrates awareness and planning to address "why?" questions through pairing with qualitative UX research.

- Considers questions of inclusion, diversity, and accessibility when planning research and interpreting results.

- Proactively considers and addresses questions of research ethics. Is sensitive to common concerns in research ethics, such as privacy.

- Ensures that research plans are actionable for UX, engineering, and business stakeholders.

- Participates in community activities or events for UX researchers or UXers generally, such as external UX conferences or internal cross-company events.

- Has two years or more experience working as a researcher in a UX organization.

- Has completed graduate training or a graduate degree in psychology, statistics, human-computer interaction, social science, or another field related to human subjects research.

- Demonstrates skill in qualitative UX research alongside quantitative research.

Skill: Statistics (Chapter 5)

Minimum

- Solid understanding of descriptive statistics including appropriate reporting for various data types, measures of central tendency, and quantiles.

- Able to import data into a common statistical computation environment for analysis.

- Uses confidence intervals and other measures of uncertainty where appropriate.

- Describes basic inferential tests correctly and fluently (e.g., t-test).

- Understands the assumptions and limitations of traditional "statistical significance."

- Demonstrates awareness of sample bias and problems of generalization.

Moderate to Advanced

- Considers and addresses threats to the reliability and validity of statistical inferences.

- Has a strong grasp of general linear models, their assumptions, and limitations.

- Designs statistical analyses with appropriate controls for skewed data, biased samples, and multivariable interactions.

- Demonstrates expertise in statistics for one or more of the following research methods: A/B testing, logs analysis, regression, survey research, clustering and classification.

- Understands and mitigates collinearity in statistical analyses.

- Uses multiple models, with appropriate model comparison statistics, when developing complex statistical analyses.

- Applies methods from machine learning along with statistics.

- Has a viewpoint and can speak to the differences between machine learning and statistics (note: there is no single right answer).

- Has completed graduate-level courses in multivariate statistics.

- Has developed novel statistical analysis methods.

- Skills in statistics approach those of working statisticians.

Skill: Programming (Chapter 6)

Minimum

- Experience with any general-purpose programming language commonly used for statistical analysis (R, Python, Julia, MATLAB, etc. SQL and SPSS do not meet this "general purpose" requirement.)

- Writes logically correct code or pseudocode to solve role-appropriate problems related to data handling and analysis. (Note: immediate syntactically correct code is not required; the focus is on logical correctness.)

- Uses functions and control structures (such as loops) appropriately.

- Clarifies assumptions about data, output, and boundary conditions.

- Gives attention to test cases to ensure correctness.

- Can write code to run start-to-finish as a complete script, without interaction.

Moderate to Advanced

- Develops code starting with the high-level structure, and then gradually addresses individual code sections.

- Thinks in terms of separable functions and writes code that is free from side effects.

- Can design logically correct code to merge data files (such as SQL JOIN commands, but may be in any language).

- Uses SQL for data importing and merging.

- Has developed or contributed to multiple projects, writing thousands of lines of code in each one.

- Has intermediate or higher fluency in multiple programming languages.

- Proactively considers edge cases and handles some of them in code.

- Designs unit tests for code.

- Correctly identifies the algorithmic complexity of realistic data processing or numeric computation problems.

- Participates in code reviews with other analysts.

- Uses collaborative coding tools, such as Git, GitHub, Microsoft Visual Studio, or proprietary platforms.

- Has developed and released applications such as a code-based website or end-user mobile or desktop application.

- Has published a library of open source code.

- Has coursework or a degree in computer science.

- Has worked as a software developer.

- Skills in programming approach those of working software engineers.

Skill: Research Communication (Chapters 4 and 13)
Minimum

- Clarifies stakeholder questions to make them suitable for research.

- Proactively inquires about how research results will be used.

- Presents research projects with clear explanations of the need for the research, methods used, data, results, recommendations, and limitations.

- Answers questions clearly and nondefensively about research design, sample validity, and limitations.

Moderate to Advanced

- Has two or more years of experience presenting UX research to industry stakeholders.

- Designs presentations to stand alone and to be readable and understandable for any stakeholder audience.

- Writes clearly and succinctly.

- Uses clear data visualizations to highlight key findings.

- Carefully separates stakeholder-friendly results and recommendations from technical details. Ensures that primary documents remain focused on action, while details are available for technical colleagues.

- Works backward to imagine presentations of results, and designs research and analyses to deliver those presentations.

- Maintains a curated research archive of reports and presentations in their product area.

- Gives attention to accessibility; designs presentations for audiences with visual or other impairments.

- Has presented research externally to sizable audiences, such as conferences.

- Is a recognized leader in research communication as demonstrated in publications, invited presentations, social media, or other venues.

- Has received awards for research presentations or publications.

APPENDIX C

References

This appendix lists each reference in the book. Entries are in alphabetical order by the author's name (or authors' names, in the case of multiple authors), and then by date.

The number preceding each reference is used in the text to identify the specific entry. For instance, the citation "See [109]" refers to the entry for the R Core Team, authors of *R: A Language and Environment for Statistical Computing*.

1. Aliannejadi, M., Zamani, H., Crestani, F., & Croft, W. B. (2021). Context-aware target apps selection and recommendation for enhancing personal mobile assistants. *ACM Transactions on Information Systems (TOIS)*, *39*(3), 1–30.

2. American Psychological Association. (2017). *Ethical Principles of Psychologists and Code of Conduct*. American Psychological Association. Retrieved from `https://www.apa.org/ethics/code/ethics-code-2017.pdf`

3. Barter, R., & Yu, B. (2017). *superheat: A Graphical Tool for Exploring Complex Datasets Using Heatmaps*. Retrieved from `https://CRAN.R-project.org/package=superheat`

4. Baxter, K., Courage, C., & Caine, K. (2015). *Understanding Your Users: A Practical Guide to User Research Methods* (2nd ed.). Morgan Kaufmann.

5. Bengtsson, H. (2022). *matrixStats: Functions that Apply to Rows and Columns of Matrices (and to Vectors)*. Retrieved from `https://CRAN.R-project.org/package=matrixStats`

6. Bernstein, G. (2021). *Research Practice: Perspectives from UX Researchers in a Changing Field*. Greggcorp.

7. Bolger, N., & Laurenceau, J.-P. (2013). *Intensive Longitudinal Methods: An Introduction to Diary and Experience Sampling Research*. Guilford Press.

8. Bostock, M., Rodden, K., Warne, K., & Russell, K. (2021). *sunburstR: Sunburst "Htmlwidget"*. Retrieved from `https://CRAN.R-project.org/package=sunburstR`

9. Bottomley, L. (1995). EPA-HTTP. Retrieved from `https://ita.ee.lbl.gov/html/contrib/EPA-HTTP.html`

10. Breiman, L. (2001). Statistical modeling: The two cultures. *Statistical Science, 16*(3), 199–231.

11. Bruce, P., Bruce, A., & Gedeck, P. (2020). *Practical Statistics for Data Scientists: 50+ Essential Concepts Using R and Python* (2nd ed.). O'Reilly Media.

12. Callegaro, M., Manfreda, K. L., & Vehovar, V. (2015). *Web Survey Methodology*. Sage.

13. Card, S., Moran, T. P., & Newell, A. (1983). *The Psychology of Human Computer Interaction*. Lawrence Erlbaum Associates.

14. Carver, R. (1978). The case against statistical significance testing. *Harvard Educational Review, 48*(3), 378–399.

15. Chapman, C. (2002). Designing software ethics. In *Proceedings of the Society for Philosophy in the Contemporary World (SPCW), 8th Annual Meeting*. Santa Fe, New Mexico.

16. Chapman, C. (2005). An exploration of writing acts. In *Proceedings of the Society for Philosophy in the Contemporary World (SPCW) 11th Annual Meeting*. Cullowhee, North Carolina.

17. Chapman, C. (2005). Software user research: Psychologist in the software industry. In R. D. Morgan, T. L. Kuther, & C. J. Habben (Eds.), *Life After Graduate School in Psychology: Insider's Advice from New Psychologists* (pp. 211–225). Psychology Press. Retrieved from `https://bit.ly/3ykMhIO`

18. Chapman, C. (2006). Exploration of a contractarian procedure for participatory design. In *Proceedings of the Society for Philosophy in the Contemporary World (SPCW) 12th Annual Meeting.* Cullowhee, North Carolina.

19. Chapman, C. (2006). Fundamental ethics in information systems. In *Proceedings of the 39th Annual Hawaii International Conference on System Sciences (HICSS 2006).* IEEE.

20. Chapman, C. (2013). 9 things clients get wrong about conjoint analysis. In B. K. Orme (Ed.), *Proceedings of the 2013 Sawtooth Software Conference* (pp. 1–11). Dana Point, CA: Sawtooth Software.

21. Chapman, C. (2020). Mind your text in marketing practice. *Journal of Marketing, 84*(1), 26–31.

22. Chapman, C., Alford, J. L., & Love, E. (2009). Exploring the reliability and validity of conjoint analysis studies. In *Proceedings of the Advanced Research Techniques (ART) Forum 2009.* Retrieved from https://drive.google.com/file/d/1Opmiji7U_Vw69q673Nz7ZkRMh4mS6P9v/view?usp=sharing

23. Chapman, C., Bahna, E., Alford, J. L., & Ellis, S. (2022). *choicetools: Tools for Choice Modeling, Conjoint Analysis, and MaxDiff Analysis of Best-Worst Surveys* (GitHub R package 0.0.0.9081.). Retrieved from https://github.com/cnchapman/choicetools

24. Chapman, C., & Callegaro, M. (2022). Kano analysis: A critical survey science review. In B. K. Orme (Ed.), *Proceedings of the 2022 Sawtooth Software Conference.* Orlando, FL.

25. Chapman, C., & Feit, E. M. (2019). *R for Marketing Research and Analytics* (2nd ed.) Springer.

26. Chapman, C., Lahav, M., & Burgess, S. (2009). Digital pen: Four rounds of ethnographic and field research. In *Proceedings of the 42nd Hawaii International Conference on System Sciences (HICSS 2009).* IEEE.

27. Chapman, C., & Love, E. (2006). Marketing research and UX research. Personal communication.

28. Chapman, C., Love, E., & Alford, J. L. (2008). Quantitative early-phase user research methods: Hard data for initial product design. In *Proceedings of the 41st Annual Hawaii International Conference on System Sciences (HICSS 2008)*. IEEE.

29. Chapman, C., & Milham, R. (2006). The personas' new clothes: Methodological and practical arguments against a popular method. In *Proceedings of the Human Factors and Ergonomics Society 50th Annual Meeting*. San Francisco, CA: Human Factors; Ergonomics Society (HFES).

30. Chapman, C., Xu, K. Z., Callegaro, M., Gao, F., & Cipollone, M. (Eds.). (2022). *Proceedings of the 2022 Quantitative User Experience Conference (QuantUXCon 2022)*. Quantitative UX Association.

31. Chihara, L. M., & Hesterberg, T. C. (2019). *Mathematical Statistics with Resampling and R* (2nd ed.) Wiley.

32. Christian, B. (2020). *The Alignment Problem: Machine Learning and Human Values*. WW Norton & Company.

33. Chrzan, K., & Orme, B. K. (2019). *Applied MaxDiff: A Practitioner's Guide to Best-Worst Scaling*. Sawtooth Software.

34. Conway, D. (2010). The data science Venn diagram. Retrieved from `http://drewconway.com/zia/2013/3/26/the-data-science-venn-diagram`

35. Croissant, Y. (2018). *mlogit: Multinomial Logit Models*. Retrieved from `https://CRAN.R-project.org/package=mlogit`

36. Croll, A., & Yoskovitz, B. (2013). *Lean Analytics: Use Data to Build a Better Startup Faster*. O'Reilly.

37. Crumlish, C. (2022). *Product Management for UX People: From Designing to Thriving in a Product World*. Rosenfeld Media.

38. Cunningham, S. (2021). *Causal Inference: The Mixtape.* Yale University Press.

39. DeVellis, R. F. (2016). *Scale Development: Theory and Applications* (4th ed.). SAGE Publications.

40. DiCiccio, T. J., & Efron, B. (1998). Bootstrap confidence intervals. *Statistical Science, 11*(3), 189–228.

41. Displayr. (2022). *flipMaxDiff: MaxDiff Experimental Design and Analysis* (R package version 0.1.0.)

42. Ericsson, K. A., & Simon, H. A. (1998). How to study thinking in everyday life: Contrasting think-aloud protocols with descriptions and explanations of thinking. *Mind, Culture, and Activity, 5*(3), 178–186.

43. Everitt, B. S., Landau, S., Leese, M., & Stahl, D. (2011). *Cluster Analysis* (5th ed.) John Wiley & Sons.

44. Fan, M., Shi, S., & Truong, K. N. (2020). Practices and challenges of using think-aloud protocols in industry: An international survey. *Journal of Usability Studies, 15*(2), 85–102.

45. Franks, B. (2020). 97 *Things About Ethics Everyone in Data Science Should Know.* O'Reilly Media.

46. Furr, R. M. (2021). *Psychometrics: An Introduction* (4th ed.). SAGE Publications.

47. Gabadinho, A., Ritschard, G., Müller, N. S., & Studer, M. (2011). Analyzing and visualizing state sequences in R with TraMineR. *Journal of Statistical Software, 40*(4), 1–37. https://doi.org/10.18637/jss.v040.i04

48. Gabadinho, A., Studer, M., Müller, N. S., Bürgin, R., Fonta, P.-A., & Ritschard, G. (2022). *Trajectory Miner: A Toolbox for Exploring and Rendering Sequences* (Version 2.2-4.) Retrieved from https://cran.r-project.org/web/packages/TraMineR/index.html

49. Gelman, A. (2021). Reflections on Breiman's two cultures of statistical modeling. *Observational Studies, 7* (1).

50. Gibson, W. (1999). Interview on NPR Talk of the Nation. Radio. Retrieved from `http://www.brianstorms.com/archives/000461.html`

51. Google. (2011). Ten things we know to be true. Retrieved from `https://about.google/philosophy/`

52. Gourville, J. T. (2004). *Why Consumers Don't Buy: The Psychology of New Product Adoption*. Harvard Business School.

53. Greene, A. E. (2013). *Writing Science in Plain English*. University of Chicago Press.

54. Grinstead, C. M., & Snell, J. L. (1997). *Introduction to Probability*. American Mathematical Society.

55. Grolemund, G., & Wickham, H. (2011). Dates and times made easy with lubridate. *Journal of Statistical Software, 40*(3), 1–25. Retrieved from `https://www.jstatsoft.org/v40/i03/`

56. Hadley Wickham. (2007). Reshaping data with the reshape package. *Journal of Statistical Software, 21*(12), 1–20. `http://www.jstatsoft.org/v21/i12/`

57. Harrell, F. E. (2015). *Regression Modeling Strategies: With Applications to Linear Models, Logistic and Ordinal Regression, and Survival Analysis* (2nd ed.) Springer.

58. Helveston, J. P. (2021). Using formr to create R-powered surveys with individualized feedback. Retrieved from `https://www.jhelvy.com/talks/2021-01-21-surveys-with-formr/`

59. Hesterberg, T. (2020). *Advice for Statisticians and Data Scientists Interested in Working at Google*. Retrieved from `https://bit.ly/3Aph498`

60. Hilbe, J. M. (2009). *Logistic Regression Models*. Chapman & Hall/CRC.

61. Hope, R. M. (2022). Rmisc: Ryan Miscellaneous. Retrieved from `https://CRAN.R-project.org/package=Rmisc`

62. Ismay, C., & Kim, A. Y. (2019). *Statistical Inference via Data Science: A ModernDive into R and the Tidyverse*. CRC Press.

63. Jarrett, C. (2021). *Surveys That Work*. Rosenfeld Media.

64. JASP Team. (2022). JASP (Version 0.16.2)[Computer software]. Retrieved from https://jasp-stats.org/

65. Kaplan, S. (2019). *The 360° Corporation: From Stakeholder Trade-Offs to Transformation*. Stanford Business Books.

66. Karimi, Z., Baraani-Dastjerdi, A., Ghasem-Aghaee, N., & Wagner, S. (2016). Using personality traits to understand the influence of personality on computer programming: An empirical study. *Journal of Cases on Information Technology, 18*(1).

67. Keiningham, T. L., Cooil, B., Andreassen, T. W., & Aksoy, L. (2007). A longitudinal examination of net promoter and firm revenue growth. *Journal of Marketing, 71*(3), 39–51.

68. Kernighan, B. W., & Ritchie, D. M. (1988). *The C Programming Language* (2nd ed.) Pearson.

69. King, R., Churchill, E. F., & Tan, C. (2017). *Designing with Data: Improving the User Experience with A/B Testing*. O'Reilly.

70. Kline, R. B. (2015). *Principles and Practice of Structural Equation Modeling* (4th ed.) Guilford Press.

71. Knaflic, C. N. (2015). *Storytelling with Data: A Data Visualization Guide for Business Professionals*. Wiley.

72. Kohavi, R., Tang, D., & Xu, Y. (2020). *Trustworthy Online Controlled Experiments: A Practical Guide to A/B Testing*. Cambridge University Press.

73. Krueger, R. A., & Casey, M. A. (2014). *Focus Groups: A Practical Guide for Applied Research* (5th ed.) SAGE Publications.

74. Krug, S. (2013). *Don't Make Me Think, Revisited: A Common Sense Approach to Web Usability* (3rd ed.) New Riders.

75. Kuhn, M. (2018). *caret: Classification and Regression Training*. Retrieved from https://CRAN.R-project.org/package=caret

76. Kuhn, M., & Johnson, K. (2018). *Applied Predictive Modeling* (Corrected 2nd printing.) Springer.

77. Kwartler, T. (2017). *Text Mining in Practice with R.* Wiley.

78. Ladner, S. (2014). *Practical Ethnography: A Guide to Doing Ethnography in the Private Sector.* Routledge.

79. Larson, W. (2019). *An Elegant Puzzle: Systems of Engineering Management.* Stripe Press.

80. Larson, W. (2021). *Staff Engineer: Leadership Beyond the Management Track.* Stripe Press.

81. Livingston, G. (2015). *Childlessness Falls, Family Size Grows Among Highly Educated Women.* Pew Research Center.

82. Lohr, S. L. (2022). *Sampling: Design and Analysis.* CRC Press.

83. Looney Tunes Wiki. (2022). ACME. Retrieved from `https://looneytunes.fandom.com/wiki/ACME`

84. Louviere, J. J., Flynn, T. N., & Marley, A. A. J. (2015). *Best-Worst Scaling: Theory, Methods and Applications.* Cambridge University Press.

85. Love, E., & Chapman, C. (2007). Issues and cases in user research for technology firms. In B. K. Orme (Ed.), *Proceedings of the 13th Sawtooth Software Conference* (pp. 43–50).

86. Lund, A. (2011). *User Experience Management: Essential Skills for Leading Effective UX Teams.* Elsevier.

87. Luster, J. (2022). Want help prioritizing items on your product roadmap? MaxDiff to the rescue! In C. Chapman, K. Z. Xu, M. Callegaro, F. Gao, & M. Cipollone (Eds.), *Proceedings of the 2022 Quantitative User Experience Conference (QuantUXCon 2022)* (pp. 259–277).

88. Marshall, B. H. (2022). *Data Conscience: Algorithmic Siege on Our Humanity.* Wiley.

89. Martens, D. (2022). *Data Science Ethics: Concepts, Techniques, and Cautionary Tales.* Oxford Univ. Press.

90. Matloff, N. S. (2011). *The Art of R Programming: A Tour of Statistical Software Design*. No Starch Press.

91. Maxwell, S. E., Delaney, H. D., & Kelley, K. (2018). *Designing Experiments and Analyzing Data: A Model Comparison Perspective* (3rd ed.) Routledge.

92. McCullough, B. D. (2021). *Business Experiments with R*. Wiley.

93. McDowell, G. L. (2015). *Cracking the Coding Interview: 189 Programming Questions and Solutions* (6th ed.) CareerCup.

94. McElreath, R. (2020). *Statistical Rethinking: A Bayesian Course with Examples in R and STAN* (2nd ed.) CRC Press.

95. Müller, H., & Sedley, A. (2014). HaTS: Large-scale in-product measurement of user attitudes & experiences with happiness tracking surveys. In *Proceedings of the 26th Australian Computer-Human Interaction Conference (OzCHI 2014)* (pp. 308–315). New York, NY, USA. Retrieved from `https://doi.org/10.1145/2686612.2686656`

96. Navarro, D. J., Foxcroft, D. R., & Faulkenberry, T. J. (2019). *Learning Statistics with JASP*: A Tutorial for Psychology Students and Other Beginners. Retrieved from `https://learnstatswithjasp.com`

97. Neuwirth, E. (2022). *RColorBrewer: ColorBrewer Palettes*. Retrieved from `https://CRAN.R-project.org/package=RColorBrewer`

98. Newton, J. D. (1987). *Uncommon Friends: Life with Thomas Edison, Henry Ford, Harvey Firestone, Alexis Carrel and Charles Lindbergh*. Harcourt Brace Jovanovich.

99. Nielsen, J. (1993). *Usability Engineering*. Morgan Kaufmann.

100. Nielsen, J. (2012). Thinking aloud: The #1 usability tool. Retrieved from `https://www.nngroup.com/articles/thinking-aloud-the-1-usability-tool/`

101. Norman, D. (2013). *The Design of Everyday Things* (Revised and expanded.) Basic Books.

102. Norton, D. (2020). *Escape Velocity: Better Metrics for Agile Teams.* Onbelay.

103. Omiwale, O. E. (2022). Making a Sankey diagram. Retrieved from `https://rpubs.com/oomiwale1/926103`

104. Orme, B. K. (Ed.). (2012). *Latent Class Manual* (version 4.5). Sawtooth Software.

105. Orme, B. K. (2019). *Sparse, Express, Bandit, Relevant Items, Tournament, Augmented, and Anchored MaxDiff: Making Sense of All Those MaxDiffs!* Sawtooth Software. Retrieved from `https://sawtoothsoftware.com/resources/technical-papers/sparse-express-bandit-relevant-items-tournament-augmented-and-anchored-maxdiff-making-sense-of-all-those-maxdiffs`

106. Orme, B. K. (2019). *Getting Started with Conjoint Analysis: Strategies for Product Design and Pricing Research* (4th ed.) Research Publishers.

107. Patten, E. (2016). *The Nation's Latino Population Is Defined by Its Youth.* Pew Research Center.

108. Portigal, S. (2013). *Interviewing Users: How to Uncover Compelling Insights.* Rosenfeld Media.

109. R Core Team. (2022). *R: A Language and Environment for Statistical Computing.* Vienna, Austria: R Foundation for Statistical Computing. Retrieved from `https://www.R-project.org/`

110. Ramaswamy, V., & Cohen, S. H. (2007). Latent class models for conjoint analysis. In Gufstafsson, A., Herrmann, A., & Huber, F. (Eds.). (2007). *Conjoint Measurement: Methods and Applications.* (pp. 295–319). Springer.

111. Reichheld, F. F. (2003). The one number you need to grow. *Harvard Business Review, 81*(12), 46–55.

112. Reilly, T. (2022). *The Staff Engineer's Path: A Guide for Individual Contributors Navigating Growth and Change.* O'Reilly Media.

113. Revilla, M. A., Saris, W. E., & Krosnick, J. A. (2014). Choosing the number of categories in agree–disagree scales. *Sociological Methods & Research, 43*(1), 73–97. `https://doi.org/10.1177/0049124113509605`

114. Robinson, E., & Nolis, J. (2020). *Build a Career in Data Science.* Manning.

115. Rodden, K. (2014). Applying a sunburst visualization to summarize user navigation sequences. *IEEE Computer Graphics and Applications, 34*(5), 50–54.

116. Rodden, K. (2020). Why use a radial data visualization? Retrieved from `https://observablehq.com/@observablehq/why-use-a-radial-data-visualization`

117. Rodden, K. (2020). Sequences sunburst. Retrieved from `https://observablehq.com/@kerryrodden/sequences-sunburst`

118. Rodden, K., Hutchinson, H., & Fu, X. (2010). Measuring the user experience on a large scale: User-centered metrics for web applications. In *Proceedings of the SIGCHI Conference on Human Factors in Computing Systems (CHI '10)* (pp. 2395–2398). Association for Computing Machinery. `https://doi.org/10.1145/1753326.1753687`

119. Rodden, K., & Leggett, M. (2010). Best of both worlds: Improving Gmail labels with the affordances of folders. In *Extended Abstracts of the SIGCHI Conference on Human Factors in Computing Systems (CHI '10)* (pp. 4587–4596). Association for Computing Machinery. `https://doi.org/10.1145/1753846.1754199`

120. Rossi, P. E., Allenby, G. M., & McCulloch, R. E. (2005). *Bayesian Statistics and Marketing.* John Wiley & Sons.

121. Sarrias, M., & Daziano, R. (2017). Multinomial logit models with continuous and discrete individual heterogeneity in R: The gmnl package. *Journal of Statistical Software, 79*(2), 1–46. `https://doi.org/10.18637/jss.v079.i02`

122. SAS Institute. (2022). *JMP 17 Design of Experiments Guide*. SAS Institute, Inc.

123. Sauro, J. (2018). Is the net promoter score a better measure than satisfaction? Retrieved from `https://measuringu.com/nps-sat/`

124. Sauro, J., & Lewis, J. R. (2016). *Quantifying the User Experience: Practical Statistics for User Research*. Morgan Kaufmann.

125. Sawtooth Software. (2022). *Past Conference Papers (1987–2022)*. Retrieved from `https://sawtoothsoftware.com/resources/ events/conferences`

126. Sawtooth Software. (2022). Export settings (MaxDiff), Lighthouse Studio Manual. Retrieved from `https://sawtoothsoftware.com/ help/lighthouse-studio/manual/analysis-manager-maxdiff- export-settings.html`

127. Schwarz, J., Chapman, C., & Feit, E. M. (2020). *Python for Marketing Research and Analytics*. Springer.

128. Sedgewick, R., & Wayne, K. (2011). *Algorithms* (4th ed.) Addison-Wesley.

129. Sermas, R. (2012). *ChoiceModelR: Choice Modeling in R*. Retrieved from `https://CRAN.R-project.org/package=ChoiceModelR`

130. Seung, T. K. (1990). Settlers vs. Explorers in philosophy (graduate seminar discussion, attended by C. Chapman). Personal communication.

131. Sher, B. (2007). *Refuse to Choose!: Use All of Your Interests, Passions, and Hobbies to Create the Life and Career of Your Dreams*. Rodale Books.

132. Silge, J., & Robinson, D. (2017). *Text Mining with R: A Tidy Approach*. O'Reilly Media.

133. Silvia, P. J., & Cotter, K. N. (2021). *Researching Daily Life: A Guide to Experience Sampling and Daily Diary Methods*. American Psychological Association.

134. Simon, H. A. (1978). Rationality as process and product of thought. *American Economic Review, 68,* 1–16.

135. Slivkins, A. (2019). Introduction to multi-armed bandits. *Foundations and Trends in Machine Learning, 12*(1-2), 1–286. https://doi.org/10.1561/2200000068

136. Spool, J. M. (2017). Net promoter score considered harmful (and what UX professionals can do about it). Retrieved from https://articles.uie.com/net-promoter-score-considered-harmful-and-what-ux-professionals-can-do-about-it/

137. Strathern, M. (1997). "Improving ratings": Audit in the British university system. *European Review, 5*(3), 305–321.

138. Sweigart, A. (2020). *Beyond the Basic Stuff with Python: Best Practices for Writing Clean Code.* No Starch Press.

139. Tansey, C. (2023). What is "stack ranking" and why is it a problem? *Lattice Magazine* Retrieved from https://lattice.com/library/what-is-stack-ranking-and-why-is-it-a-problem.

140. Train, K. E. (2009). *Discrete Choice Methods with Simulation.* Cambridge Univ. Press.

141. Travis, D., & Hodgson, P. (2019). *Think Like a UX Researcher.* CRC Press.

142. Tsitoara, M. (2019). *Beginning Git and GitHub: A Comprehensive Guide to Version Control, Project Management, and Teamwork for the New Developer.* Apress.

143. Tukey, J. W. (1986). Sunset salvo. *The American Statistician, 40*(1), 72–76. Retrieved from https://www.jstor.org/stable/2683137

144. Tullis, T., & Albert, W. (2022). *Measuring the User Experience: Collecting, Analyzing, and Presenting Usability Metrics* (3rd ed.) Morgan Kaufmann.

145. Wainwright, K., & Remy, L. (2022). Your surveys aren't accessible. In C. Chapman, K. Z. Xu, M. Callegaro, F. Gao, & M. Cipollone

(Eds.), *Proceedings of the 2022 Quantitative User Experience Conference (QuantUXCon 2022)*.

146.   Wang, B., Wu, P., Kwan, B., Tu, X., & Feng, C. (2018). Simpson's paradox: examples. *Shanghai Archives of Psychiatry, 30*(2), 139–143.

147.   Wei, T., & Simko, V. (2021). *R package corrplot: Visualization of a Correlation Matrix*. Retrieved from `https://github.com/taiyun/corrplot`

148.   Wickham, H. (2016). *ggplot2: Elegant Graphics for Data Analysis* (2nd ed.) Springer.

149.   Wickham, H., & Grolemund, G. (2017). *R for Data Science*. O'Reilly Media.

150.   Wickham, H., Hester, J., Chang, W., & Bryan, J. (2022). *devtools: Tools to Make Developing R Packages Easier*. Retrieved from `https://CRAN.R-project.org/package=devtools`

151.   Wickham, H., & Seidel, D. (2022). *scales: Scale Functions for Visualization*. Retrieved from `https://CRAN.R-project.org/package=scales`

152.   Wilke, C. O. (2022). *ggridges: Ridgeline Plots in ggplot2*. Retrieved from `https://CRAN.R-project.org/package=ggridges`

153.   Wirth, N. (1976). *Algorithms + Data Structures = Programs*. Prentice-Hall.

Index

A

B

C

© Chris Chapman and Kerry Rodden 2023
C. Chapman and K. Rodden, *Quantitative User Experience Research*,
https://doi.org/10.1007/978-1-4842-9268-6

Printed in the United States
by Baker & Taylor Publisher Services